Raincoast Sasquatch
THE BIGFOOT/SASQUATCH RECORDS OF SOUTHEAST ALASKA, COASTAL BRITISH COLUMBIA, & NORTHWEST WASHINGTON, FROM PUGET SOUND TO YAKUTAT

J. Robert Alley

ISBN-13: 978-0-88839-143-8 *[2018 B&W Reprint]*
ISBN-13: 978-0-88839-508-5 *[2003 1st Ed. Colour]*
Copyright © 2003 J. Robert Alley
Third printing 2018 (B&W Edition)

Cataloging in Publication Data

Alley, J. Robert
  Raincoast sasquatch : the bigfoot/sasquatch records of southeast Alaska, coastal British Columbia, & northwest Washington, from Puget Sound to Yakutat / J. Robert Alley.
  Includes bibliographical references and index.
  ISBN 0-88839-508-6 [2003]   ISBN 0-88839-143-9 [2018]
  1. Sasquatch—Northwest, Pacific. I. Title.
  QL89.2.S2A44 2003   001.944   C2003-910446-X

All rights reserved. No part of this publication may be reproduced, stored in a retrieval system or transmitted, in any form or by any means, electronic, mechanical, photocopying, recording, or otherwise, without the prior written permission of Hancock House Publishers.

Printed in the USA

Editor: Nancy Miller
Production: Irene Hannestad
Cover design: Ingrid Luters
Front cover photos: Northwest Coast mask, American Museum of Natural History
Back cover photos: J. Robert Alley
All photos and illustrations by J. Robert Alley unless otherwise credited.

We acknowledge the financial support of the Government of Canada through the Book Publishing Industry Development Program (BPIDP) for our publishing activities.

Crypto Editions is an imprint of Hancock House Publishers

Published simultaneously in Canada and the United States by

HANCOCK HOUSE PUBLISHERS LTD.
19313 Zero Avenue, Surrey, B.C. Canada V3S 9R9
(604) 538-1114 Fax (604) 538-2262

HANCOCK HOUSE PUBLISHERS
1431 Harrison Avenue, Blaine, WA U.S.A. 98230-5005
(604) 538-1114 Fax (604) 538-2262

Website: www.hancockhouse.com
Email: sales@hancockhouse.com

Contents

Acknowledgments 4
Preface 5
Introduction 9

Part One / A Good Look at the Sasquatch
1. Along the Roads 17
2. North Up the Panhandle 33
3. Close Encounters 39
4. Aquatic Anthropoids 57
5. Trouble with Sasquatches 65
6. Sasquatches on the Shore 71
7. Chasing Deer and Other Stories 89
8. A Sightings Digest 94
9. Old Stories from the B.C./Alaska Coast 98
10. Native Sasquatch Folklore 137

Part Two / Circumstantial Evidence
11. Big Footprints in the Alaskan Panhandle 167
12. Screams in the Forest 191
13. Strange Encounters in Southeast Alaska 215
14. Possible Sasquatch Nests and Other Signs ... 237
15. Traces of Hair and "Missing Bones" 259

Part Three / What Could It Be?
16. North American Ape or Relic Hominid? 271
17. Collecting a Sasquatch & Final Thoughts 283

Appendix One Field Guide to the Pacific Coastal Sasquatch 293
Appendix Two Summary List of Southeast Alaskan Reports 305
Appendix Three Summary List of B.C. Coastal Reports 318
Maps Sighting Report Locations 331

Bibliography 347
Index 352

Acknowledgments

The author wishes to especially thank John Green, Larry Kaniut, Al Jackson, Eric and Leif Seivertson, Dr. W.H. Fahrenbach, Dr. Jeff Meldrum, Bobbie Short, Ray Crowe, Virginia Colp and Dr. John Bindernagel for their knowledgeable contributions, suggestions and encouragement. Bravo to all those individuals who have come forward with reports to be recorded and scrutinized. Errors or omissions are the author's.

Illustration: Patrick Beaton

Readers wishing to contact the author with coastal sasquatch reports may email Dr. Robert Alley <ralley@kpunet.net>.

Preface

In the 1960s at the University of Manitoba, while preparing for a career in the health sciences, I took courses in biology and anthropology. This included zoology and physical anthropology on the one hand and courses in cultural anthropology and Native ethnography on the other. Around that time, I began to see various Canadian newspaper reports of "sasquatches." I wasn't quite sure what a sasquatch was, so I bit my lip and opened a dictionary, not really expecting to find anything at all—but I was wrong. The sasquatch is, to use the definition from the 1974 college edition of *Webster's Dictionary*, "a huge, hairy, manlike creature with long arms, reputed to live in the mountains of Northwest North America."

"Right...sure thing," I said to myself, pushing the dictionary away and, with it, notions of gorillas swaggering through the giant spruce and cedar of the West Coast. But about a month later in the anthropology lab, going over some human and primate bones, I was struck by certain similarities between several larger fossil primates and the descriptions of the sasquatch from both folklore and present-day reports.

One of my anthropology professors, who had PhDs in both physical and cultural anthropology, suggested that the possibility of a real creature was certainly there. My own sasquatch investigation truly started in 1970–71 while doing research for an anthropology thesis at the university. Originally, my research was bibliographic, simply a comparative study of Native folklore across Canada, focusing on the characteristics and behavior of sasquatchlike creatures in old ethnographies. Graduating in anthropology in 1971, I closed my books on sasquatch folklore, but made a mental note to keep an open mind on the subject.

As I continued with studies at the medical campus in Winnipeg and pursued a career in physical therapy, I had little time for newspaper reports of wild hominids from the Pacific Northwest, northern Manitoba or elsewhere. It was not until I moved to the western coast of Canada in the 1970s and met Native patients there that I would again hear much about sasquatches. Working as a physical therapy intern and traveling throughout British Columbia, I heard quite a few first-hand accounts. While I continued to be somewhat skeptical of individual reports, I recalled the considered opinion of several of my former professors, who had reminded us that there could possibly be just such a creature waiting for another culture to "discover" it.

Then, on a backpacking vacation in the Rocky Mountain foothills near Nordegg, Alberta, in August, 1973, I found a sixteen-inch track in mud, showing five toes, with no claw prints. The first toe was the largest and foremost—not belonging to any bear—and larger than any human foot I had ever seen. I was quite familiar with the variations in human anatomy, tracks and gait, and I did not think it likely that any huge human with a pituitary disorder was running around the Rocky Mountain foothills barefoot. Because I was aware of two other large mammals that were undiscovered in western Canada until recent times (the wood bison of northern Alberta and the cougar of eastern Manitoba), I privately conceded the hypothetical existence of another species as yet undiscovered—the sasquatch.

Taking my observations to Dr. Vladimir Markotic (now deceased), a professor of anthropology at the University of Calgary, I asked him what thoughts he had on the subject. The professor sat me down and said, "What you have discovered by now should confirm your suspicions that there are, quite probably, numerous living hominids that have not met the formal criteria for inclusion in the lists of known species. It should not be surprising in the least! What exactly sasquatches are is something that remains for us to determine."

In 1974, I took a hospital position and moved to Nanaimo, on Vancouver Island, British Columbia. That same year, I took a ferry to meet with authors and sasquatch researchers John Green and Rene Dahinden at their homes in B.C.'s Fraser Valley. I was surprised by the number of large tracks both men had cast in plaster and researched. Together they must have had 100. Green, especially, was (and still is) a skilled interviewer and journalist. He is a graduate of Columbia University, a longtime newspaper editor and a man not easily taken in by anyone. His database of sighting reports he considered valid, after culling out the fabrications and mistaken identifications, was astounding. Dahinden was a hard-boiled field researcher. Outwardly, he was cynical and tough, but good-hearted enough to take the time, over several trips, to share with me how to distinguish viable sasquatch tracks from hoaxes. Where Rene was taciturn, Green would share data and examine hypotheses with scientist and lay person alike. Neither man was very enthusiastic about Native folklore, although Green had certainly listened to his share of reports, some with a "supernatural" twist to them. Both men had no doubt as to the existence of a sizable population of sasquatches along the West Coast and in the interior. As well, they had similar ideas about the nature and behavior of the sasquatch. They both felt then that some hard evidence to satisfy the biologists was what was really required and, if possible, a few scientists willing to examine that evidence.

In view of the ridicule, absence of funding and stolid skepticism that had got in the way of any sustained formal research on sasquatches, I had to agree with them. But back "over on the island" and "up the coast," people didn't seem to worry too much about whether the scientists were satisfied or not. Many Natives, whom I would later come to know as patients in B.C., Alaska and Washington, repeatedly insisted the creature was real by anyone's standards and very much alive along the West Coast. To them at least, it was in many ways well known. For me, their personal accounts began to correspond very clearly with much of the earlier folklore and modern newspaper reports as well. It seemed peculiar indeed that the descriptions of such creatures should vary so little between cultures and over time. What emerges from an examination of all these perspectives is truly amazing. It is the picture of a living, breathing species of hominid, unlike any known primate today, as different from known species of fossil man as from any living great ape.

This book is a collection of a good many years of interviewing witnesses, ethnographic research and field research involving various types of evidence. While today sasquatch/bigfoot reports have continued to emanate from about every large forest, park or wilderness in North America, it seemed to me somehow fitting to search for the sasquatch in the rain-drenched coastal forests, far from man and the noise of his cities and roads. The pages that follow focus on the surprising existence of a large, reclusive species of relic hominid along the remote western coasts of British Columbia and southeast Alaska. I should mention that despite some variety of descriptions and names given by witnesses in depicting what they have stated they have seen, e.g., hairy man, creature, etc., I have chosen to refer to such hair-covered hominids generically as sasquatches. The word hominid refers to the family of man, all two-legged forms, past and present. Where hominoid is used, it refers to both the great apes and man. Physical anthropologists may point out that there may hypothetically be more than one species of hominid involved in these reported sightings. While I do not disagree with them that future research is required to confirm or dispute this, it is not any more relevant at this point than the differences between such creatures in the various Native historical belief systems. It should also be noted that when I have transcribed people's stories or used quotes from written material, I have left the grammar, speech patterns, etc., as they were originally. While this can result in some discrepancy in style consistency, I believe this is important as it not only maintains legitimacy, it also helps to capture the emotions and experiences of the witnesses. I have inserted my comments in square brackets when I feel there is a need for clarification. Unless otherwise noted, all the photographs in the book were taken by me.

Introduction

The word "sasquatch" was first used in North American media in the 1920s by J. W. Burns,[1] a British Columbia teacher from Agassiz, in the lower Fraser Valley. Burns introduced the Anglicized name sasquatch, based on a Native Coast Salish name for a large, hair-covered, manlike creature, to North American readers in an article for the national Canadian news magazine *MacLean's* in 1929. She popularized the word in an item titled, "Introducing B.C.'s Hairy Giants—A collection of strange tales about British Columbia's wild men as told by those who say they have seen them." Recently, Native ethnographer Gail Highpine[2] listed more than 200 different names for sasquatchlike creatures among Native North American tribes. In the lower Fraser Valley, among the Coast Salish, it is *sesxech*. Over much of southeastern Alaska it is *kushtakaa*, or on southern Prince of Wales Island, *gagiit*. To the Tsimshean of coastal British Columbia and Alaska's Annette Island it is *ba'oosh* or *ba'wis*. Such were the names for hair-covered, manlike creatures as given by Native people when the explorers and colonists arrived, and so they persist today.

Regardless of subtle differences, all such hairy, upright hominoids are, for collective purposes here, generally termed sasquatches. Since the mid-1900s, there has come to be a growing body of newspaper and other reports from the West Coast of just such a species of North American primate. It has since occurred to many that the legends and reports might in some way be related.

Zoologists have been commenting on sasquatch since the middle of the last century. In 1955, a Belgian cryptozoolgist specializing in researching reports of new or unidentified species, Dr. Bernard Heuvelmans, published a book detailing his research. In it he dealt with the hypothetical existence of creatures such as the Yeti, the giant squid (then unproven) and the sasquatch. The abridged English translation of his book, *On the Track of Unknown Animals*, published in 1958, included a careful collection and analysis of sasquatch reports.

In 1961, zoologist, animal collector and author Ivan T. Sanderson published a remarkable book entitled *Abominable Snowmen: Legend Come to Life*. Sanderson offered widespread documented evidence suggesting the possible existence of at least four different species of hair-covered, upright primates in forests around the world—one being the sasquatch of the Pacific Northwest. In his book, Sanderson related various documented sightings and tracks dating from the 1800s onward and samples of traditional Native stories from the coastal mountains and forests of the Pacific Northwest. Sanderson stated succinctly: "I went aside in British Columbia to investigate their long-renowned Sasquatch, only to find it was just as definite, and apparently identical to these Oh-mahs (or 'Bigfeet') of California. Subsequent research has, what is more, brought to light a mass of other reports of similar things from Quebec, the Canadian Northwest Territories, the Yukon, the Idaho Rockies, Washington, and Oregon."[3] He concluded that the results suggested a species of nocturnal apelike, yet somewhat manlike, creature, still uncollected by any zoologist. It is worth noting that Sanderson, an upstanding member of the Royal Zoological Society, a graduate of Oxford University in zoology and a well-regarded author and collector of unique species worldwide, would risk the inevitable loss of credibility among his fellow scientists.

British Columbia newspaperman John Green of Harrison Hot Springs, B.C., reviewed Burn's earlier work, and continued collecting both old and new reports, publishing *On the Track of the Sasquatch* in 1968. Initially focusing on British Columbia, Green expanded a collection of thousands of North American reports that he has subjected to investigation, indexing and analysis. Many of these he published in 1973 under the title *The Sasquatch File*. Along with the late veteran field investigators Rene Dahinden and Bob Titmus, among others, he personally managed to interview a great number of individual witnesses and traveled extensively in the northwest to do so, often casting tracks. Green has, from the outset, selflessly shared the data, track casts and photographs with researchers.

Using computer facilities at the University of British Columbia to evaluate sasquatch report data, he assembled a database, comprising close to 3,000 North American sighting reports. These were collected over decades from hunters, police, fish and wildlife officers, judges, teachers and, in fact, every manner of person who had nothing to gain and perhaps everything to lose from coming forward. But now a growing number of scientists are openly suggesting that a real North American primate is ready for serious study. They are saying that even without a conventional specimen, the evidence of sightings, tracks, film, photographs and hair is compelling.

Veteran Canadian wildlife biologist Dr. John Bindernagel has recently summed up this need for a whole new perspective: "As a wildlife biologist I have approached the existence of the sasquatch in the same way I would assess the existence of any large mammal, be it the grizzly bear, black bear or mountain gorilla. In short, my interest in the sasquatch begins at that point in the discussion when a skeptic finally asks, 'If it does exist, what does it eat? How does it behave? And how does it survive the winters?' Having accepted the available evidence as sufficient to document the existence of the sasquatch, I feel that *these* are the questions we should now be addressing."[4]

What may be most compelling are the reports themselves. They describe in detail a creature that is occasionally glimpsed and only rarely observed. Together they form a body of anecdotal evidence suggesting that, as usual with the biological sciences, the list of known species in North America is not complete.

Folklore, Native masks and totems by themselves do not in any way prove that there is a scattered population of nocturnal upright apes or fossil men strolling unnoticed through the forests of North America. It is worth noting, however, that a few anthropologists and biologists have in the past decade begun to look at a growing number of hair samples, "fingerprints" from tracks and the hundreds, perhaps now thousands, of plaster casts of tracks, taped recordings of alleged sasquatch vocalizations and so on. A mere legend cannot leave tracks or hair, and it doesn't jump easily from one culture to another, Native or otherwise. It appears to many people that some form of large primate out there has been doing both, and not without being seen.

Science is very conservative. Fewer than a dozen PhD holders

11

have openly dared to be the first to define sasquatches as mammals in the fauna of North America. One such scientist is the accomplished Dr. Grover Krantz of Washington State, an eminent scholar on *Homo erectus*, a fossil form of early man. Another exception is wildlife biologist John Bindernagel of British Columbia. Although more skeptical anthropologists cite lack of evidence, they usually have other interests and favored areas of research. As well, established textbooks have already declared that although such a creature certainly did exist until recently, it must certainly by now be extinct. It is the easiest thing for them to say and the only thing they can do without risking their careers. Ridicule threatens their livelihood. None of this should suggest that evidence would be easy to obtain. Such a hypothetical animal would not be likely to stand obligingly in front of a logging truck at night or pose for any hunter with a rifle aimed at it during daytime. No researchers that I am aware of have ever been awarded a grant to find a sasquatch. Such funds are readily available for researching known species that have some value to big game hunters but little else. Environmental organizations interested in saving rare or endangered animals are not involved in researching this creature. And only animals of interest to big game management or human medical research are on the list of funded

Figure 1: Human and gorilla crania compared with that of a *Gigantopithecus*, as reconstructed by Dr. Grover Krantz. (Photo courtesy Dr. Grover Krantz, permission Hancock House Publishers)

species. In other words, there are no grants to procure a specimen of a sasquatch, because no one has produced a body to prove it is real.

In most instances, witnesses reporting sasquatches appear to have had more of a concern with forgetting the hair-raising experience and avoiding imagined legal consequences or the persistent attention of the press, than with the detached, analytic process of furthering science. What was a healthy skepticism in the sixties has, through fantastic caricatures in the tabloid media, become lazy ridicule. Even hard biological evidence presents a problem. If someone ever shoots a sasquatch one night out of Alert Bay, B.C., or Hydaburg, Alaska, we are not likely to hear about it until the ravens and bears have dispersed any possible remains (which of course takes only a few days). A forest full of voles, mice or porcupines makes short work of bones of all species, taking advantage of the bones' rich supply of phosphorus and calcium. In most cases, the complete obliteration of bone material takes only a few months. Considering the huge population of bears, wolves and cougars dying "natural" deaths on this continent each year, it's staggering how pitifully few bones of any species are actually found.

No item of circumstantial evidence presented separately, be it film, track cast or hair, could be said to satisfy all skeptics at once. There are numerous other ways to come to logical conclusions however. Much of what follows is presented for those who would wish to take a closer look at coastal reports of sasquatches, their tracks, possible vocalizations and other alleged behavior, and form their own conclusions. There are essentially three things to consider: documented reports of sasquatch sightings; their historical counterpart—native folklore; and, circumstantial evidence such as tracks, hair, unusual nests, etc.

We already know that a hominid could be as large as reported sasquatches. Physical anthropology texts all agree that one such creature, the Asian giant hominoid *Gigantopithecus*, is so far the largest and longest surviving hominid in fossil history. Is there any evidence that an ongoing subspecies or descendant of this giant creature lives on in the mountain forests of North America? Quite a bit more than might be imagined, as it turns out.

Scientific names for the maker of large hominid tracks in the Pacific Northwest, such as *Gigantopithecus canadensis* as formally proposed by Krantz,[5] have already been put forth. Some scientists have concluded on the basis of track anatomy, hair structure or fingerprint-ridge type patterns called dermatoglyphics, that we are dealing with a true primate and are getting ready to deal with the "real McCoy." In fact, a previous primatology director of the Smithsonian, the respected Dr. John Napier, a world-renowned specialist in primates, actually had the enthusiasm to publish a thoughtful and revealing book supporting the existence of such a creature, entitled *Bigfoot*.

So, why has the sasquatch gained less prominence among the lettered professors and more among the sensational tabloids thirty years later? Science is directed by a conservative few, and the rest of the population is fed the sensationalism of the popular media. Dr. Napier knew that many people are naturally uncomfortable reaching such conclusions on their own, and that the tabloids would rule the day for 99 percent of the population. People like a bogeyman. The "bigfoot" of today's tabloids is so far removed from the nature of even any real primate that it bears little resemblance to the creatures in the reports that follow. But as Napier summed up: "I am convinced that the Sasquatch exists, but whether it is all that it is cracked up to be is another matter altogether. There must be something in northwest America that needs explaining, and that something leaves manlike footprints. The evidence that I have adduced in favour of the reality of the Sasquatch is not hard evidence; few physicists, biologists or chemists would accept it, but nevertheless, it is evidence and cannot be ignored."[6]

We North Americans seem to have a hard time with any giants or wild men who aren't nasty. This may have had its origins in northern Europe with "trolls" and terrifying hairy man-beasts such as "Grendel" in the Anglo-Saxon epic *Beowulf*. In fact, a lot of people found it remarkable when the late Dr. Dian Fossey, a well-educated American woman, moved to the rain forests of eastern Africa and parked herself for two years right in the middle of a large troop of 400-pound gorillas. Fossey observed, "I have always been convinced of the intrinsically gentle nature of gorillas and felt their charges were basically bluff in nature, so I never hesitated to hold my ground. However, because of the intensity of their screams and the speed of their approaches, I found it possible to face charging gorillas only by clinging to surrounding vegetation for dear life."[7] Several years previously, an American wildlife biologist, Dr. George Schaller, spent nine months doing virtually the same thing. In spite of numerous

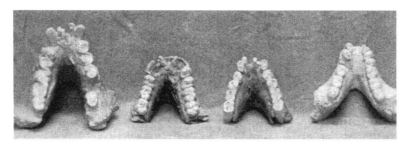

Figure 2: Four Asian fossil mandibles of *Gigantopithecus*, the largest hominoid ever known to have lived. (Photo courtesy of Dr. Grover Krantz, permission Hancock House Publishers)

bluff charges by the dominant male gorillas, both researchers emerged unscathed. The vicious ape of *King Kong* and *Tarzan* dies hard. "Gentle" is not what we've been fed. But by the analysis of sasquatch reports, shy, retiring and gentle if unprovoked are the characteristics of what wildlife biologist John Bindernagel calls "North America's most misunderstood large mammal."[8]

Whenever a rare species is "discovered," one doesn't come to know any more of its natural behavior than witnesses' reports suggest. But Green's earlier documentation dealing with sasquatch reports and analysis, as well as detailed investigation into different purported behaviors of sasquatches, has already given us a consistent picture of what people are seeing. Given the thousands of reports of a large, apparently reclusive hominid in better-explored regions of North America, I wanted to see if details of sasquatch reports from Alaska's hidden raincoast would in any way differ from those further south. My personal interest in the sasquatch is not the bigfoot hunter's question, "How can I prove to the world that it is real?" Rather, having seen sufficient evidence of all kinds over twenty-five years to suggest such a hominid survives, my question is "What kind of primate could this be?" and "How does it live and manage to remain so seldom seen?"

Alaska is not the first place that comes to mind when we hear the words bigfoot or sasquatch. British Columbia, Washington and Oregon seem more likely spots. But if something were to remain unreported, unseen, what better place than America's largest national forest, and the neighboring northern coast of British Columbia? Southeast Alaska, from Icy Bay in the north to the tip of Misty

Fjords National Monument in the south, is only 525 miles long, but it has 15,500 miles of marine shoreline. It is because these coastal rainforests are sparsely populated and difficult to access that a quiet depth of reports here should not come as a great surprise.

Watching wildlife in southeast Alaska can be difficult. Inside the thick forests one cannot see far. Short days in winter and impenetrable vegetation year-round force our species to deal with its own sensory limitations. The coastal forests are a special place where many animals exist close to man without his ever knowing. Animals living here leave little to mark their passing, and some are more often heard than observed. In this environment of more than 150 inches of annual rainfall, with life shrouded by cloud and a dark maze of green, there is much more to be discovered than at first meets the eye, as we shall see...

    1. Burns, J. W., "Introducing B.C.'s Hairy Giants," *MacLean's Magazine*, 1 April, 1929, p. 9.

    2. Highpine, Gail, personal communication, 1999.

    3. Sanderson, Ivan, 1961, *Abominable Snowmen: Legend Come to Life*, Chilton, N.Y., p. 21.

    4. Bindernagel, John, 1999, *North America's Great Ape: the Sasquatch*, Beachcomber Books, Courtenay, B.C., p. 3.

    5. Krantz, Grover S., 1999, *Bigfoot Sasquatch Evidence*, Hancock House Publishers, Surrey, B.C., appendix B.

    6. Napier, John, 1972, *Bigfoot*, E. P. Dutton and Company, New York, N.Y, p. 207.

    7. Fossey, Dian, 1983, *Gorillas in the Mist*, Houghton Mifflin Company, Boston, p. 56.

    8. Bindernagel, John, 1998, *North America's Great Ape: the Sasquatch*, Beachcomber Books, Courtenay, B.C.

Part One / A Good Look at the Sasquatch

1 Along the Roads

Alaska continues to be, outside of its three largest cities, quite provincial, and Alaskans generally tend to be hard working and inclined to be rather tight-lipped about fish runs, gold claims and practically everything else. In spite of this, there really has not been any one period since 1950 when sasquatches, their tracks or possible vocalizations have not been reported in southeast Alaska. Most of the following stories have not been made known outside of friends and family until recently. It is likely that one reason for this is the fear of ridicule. It is perhaps for this reason that some of the following witnesses have not chosen to have their full identities published. Another may be that the existence of the creature along the Alaskan coast has not been widely mentioned in the local media. It is noteworthy that with the increased media attention given to sasquatches in the U.S. and Canada in the past three decades, more than a few southeast Alaskans have come to the conclusion that what they have experienced may be something more than simply a Native folk tale.

Native southeast Alaskans have certainly always had their stories of wild creatures resembling men. It may have been easy, in the early 1900s, to assume that any such story coming from an old sourdough, fresh from the Alaskan bush, was embellished by Native influence. With modern reports, however, this is much less likely.

Before getting into reports, a word or two on sasquatch names is called for. The word *kushtakaa*, the Tlingit name for a lost human transformed into a hair-covered creature, is still used in southeast Alaska today, even by some non-Natives, as a euphemism for sasquatch. This has made the whole thing rather confusing. There are still traditional Tlingits or Haidas who firmly believe that all man-sized, hair-covered creatures are transformed humans, or *kushtakaas*, and that any larger, hair-covered hominids would have to be called sasquatches. In some of the earlier reports of sasquatchlike creatures assembled here there is often little awareness of widespread coastal sasquatch documentation. The terms sasquatch or bigfoot have been largely unused in southeast Alaska, with "hairy man," *kushtakaa* or other local names usually given. Furthermore, some local Alaskans have apparently had the notion that such hairy, manlike creatures were only to be found in southeast Alaska. Where the term *kushtakaa* has been used in some sighting reports, it is taken here to signify simply "upright, hair-covered hominid," or sasquatch, rather than the historical, transformational interpretation (to be dealt with later).

There have been a significant number of sightings along southeast Alaskan roads, by both highway motorists at night, as well as by hunters along logging roads at dawn and dusk. In most cases, both sasquatches and humans seem to have been mutually surprised, and the reported sasquatches have not seemed much interested in humans. Most sightings along roads are brief, the creature in view for less than ten seconds or so. Witnesses appear to have been even less inclined to stop and stare. Only reports from individuals found credible are documented. Each report is followed by a code containing the year, (eg 1967), and a map number following the code, eg. [S67C-map5]. What follows are some of these accounts.

Revilla' (Revillagigedo) Island Area Roadside Sightings

Mr. E. B., a retired Annette Island fisherman in his fifties, reported having seen something unusual in the bush sometime in the 1970s, while fall deer hunting five miles south of Metlakatla, Annette Island. He had been walking a gravel road into the mountains, through patches of open muskeg in hemlock and cedar forest, when he had been surprised by something running across the road about sixty yards away. This is what he said in a 1999 interview:

"It was around a curve in the road just west of Purple Mountain, near Metlakatla, but there was enough spacing in the trees to make out a light brown form running across the road from left to right. It wasn't low like a bear, and it didn't bound like a deer. I only saw it for a few seconds. It was about as tall as I am, six feet, and I have never seen anything move that fast before. I moved up the road quickly and if it was a deer running, I should have been able to have seen it. It had covered the open ground real fast and was gone. I'm not sure what it was, but I think it could have been one of those creatures." [S70C-map5]

Sasquatches are just as often reported seen along a road as by the ocean, and frequently near both. A longtime resident of Revilla' Island, Mrs. K. W., reported a typically brief roadside sighting along an undeveloped stretch of the North Tongass Highway, just north of Mile 13 and northwest of Ketchikan, Revilla' Island. In a 1999 interview, she summed up her experience:

"Late one afternoon in the late 70s, summertime, I had been driving southbound, toward Ketchikan with passengers, coming up on the scenic viewpoint that overlooks Tongass Narrows. As I came around a curve, I noticed a seven-foot, dark, hair-covered creature standing on the right-hand side of the road, just at the edge of the forest. It was on two legs and jumped quickly into the trees and was gone. But it stayed upright as it disappeared into the trees!" [S70B-map5]

The forest along the viewpoint looks now much as it did then, dark second-growth hemlock and cedar, with patches of pine and muskeg. What is curious about this minor "hot spot" location is that there are at least two other reports of sasquatches seen less than a mile away. One was a group sighting by five youths in 1955; another was the following sighting by Ketchikan cyclist Gerald P. in 1998. Three reports within one mile is unusual, but over a forty-three-year time span is downright peculiar!

Few people in southeast Alaska report seeing sasquatches in broad daylight, if they report them at all. The report that follows is unusual in that while it occurred along the same mile of highway north of Ketchikan as two other reports, it apparently occurred in the middle of the afternoon. Gerald P. is a serious young Ketchikan man who consented to being interviewed in December of 1998. In a quiet voice he recounted details of a surprising sighting he said he had while cycling back to town in August of that year. He stated he had been visiting a friend at the end of the North Tongass Highway at Settler's Cove recreation area, and in midafternoon was on his bike headed southbound down the highway. He had been cycling along the highway near mile-marker twelve, coasting quietly along at a good speed, having just finished a long downhill stretch. The sides of the highway beyond the ditches were well forested and the ditches

Figure 3: Illustration of sasquatch reported by cyclist Gerald P. on Revilla' Island Highway, 1998.

were empty of shrubs, just grass and such. Gerald said he happened to glance to his left across the other lane at a movement that caught his eye. At that moment he felt his hair go up.

"I had just slowed down from seeing how fast I could go, when I suddenly noticed something standing in the ditch on my left as I went by, less than twenty feet away. It was big and looked like it weighed maybe 275 to 300 pounds, with long arms and dark brown fur. I got a lungful of the most awful gagging smell, like a combination of rotten meat, urine and damp earth. It was opposite me in the bottom of the ditch with its head and shoulders bending down. It was shaped about like a man but larger; it would have been about seven feet tall if it had stood up straight. The hair on the body was a couple of inches long. The hair on the top of the head was longer and matted. Its head was bent over as I went by.

"It didn't look up, and I don't think it heard me coast up to where it was. I couldn't see its two legs well, just from the tops of the thighs up, but it was reaching down and looked like it was digging at something in the slope of the ditch nearest me. There were flies buzzing all around it.

"It smelled so bad it made me want to throw up. It was shaggy and not really heavily built, it kind of looked like it wasn't doing too well, kind of on the thin side. I know it wasn't a bear. I couldn't see its arms from the forearms down. I thought it might have had a dead dog it was digging at or something because of the smell. Its head was down and I couldn't see a neck or its face.

"I was scared and started pedaling real hard and heard crunching sounds like wood breaking behind me as I went past. I was scared—I felt my hair go up. I thought maybe it was coming after me, but after fifty feet or so I looked back and it was gone. I guess it went back into the woods on that side. The smell really made me want to gag. I didn't stop pedaling once till I got home in town." [S98A-map5]

Sasquatches are commonly reported on this southwest corner of Revilla' Island. There are no human habitations to speak of along the inland side of this particular stretch of highway, and the forest continues uninterrupted from there some sixty miles north across the island. Deer cross the road regularly; there are clams on the beaches below. Not a bad place to live, for a sasquatch at any rate.

Miss Robin S., a Ketchikan retail clerk, told me in 1999 of a second-hand report she and her sister had heard in late 1986, which was said to have happened at a picnic shelter just five miles south of Ketchikan. Her account went as follows:

"My boyfriend and his friend had told me about a strange thing they had seen late one night at Bugge's Beach (a half-mile north of Saxman, south of Ketchikan). It was August, and they had camped out and stayed up late at the open picnic shelter in the trees between the beach and the highway. They had been up all night, had a fire going, and it was about four or five in the morning. They were at the tables in the shelter when they heard heavy footsteps crossing the highway. They said they watched as a real tall, darkly furred, man-shaped figure came walking down the hiking trail that passes a few yards from the shelter. They said they watched as it walked right past them at their fire, less than thirty feet away, heading south on the forest trail that parallels the beach." [S86C-map5]

The year 1999 brought another sighting near Saxman, also in August, and just a half-mile from the previous spot. Saxman resident Mike V. reported that he and friends saw a creature that was "either a *kushtakaa* or a sasquatch" from his parked car.

"I had parked with friends on a dead-end service road north of Saxman one evening in August, and was enjoying some music with the other three in the car. I had the car running and the headlights aimed ahead down the blocked-off road and into the hemlock and pine on either side. There was a small utility building set into the forest, twenty yards ahead past a low log barricade, and the building had a single outdoor light on. Some time after midnight, we were all surprised to see a seven-foot, dark brown or black, upright creature walk real smooth across the road. It walked into the light about a hundred feet in front of us, left to right.

"It was swinging its arms and was heavy built. It didn't turn to look at the headlights and continued into the trees. That was it, we stayed a while but it didn't come back." The other witnesses stated they did not hear any sound or detect any odor. Mike later estimated its weight at possibly up to 500 pounds.

In parts of the Pacific Northwest, sasquatch sightings in any one spot appear to fall off after a number of years, for reasons unknown. If reports obtained herein are any indication, Ketchikan's Ward Lake, a popular forest recreation spot less than ten miles north of the city, shows no sign of doing so. A middle-aged Ketchikan grandmother, who wishes to remain anonymous, told me in an interview in March, 2000, of a sighting she had early one morning in late November, 1999. She stated that, at about 5:00 a.m., she had taken her usual morning drive out to the end of the Ward Lake Road at Signal Creek Campground to park, enjoy a cup of coffee and wait for the sunrise.

"It was dark and, because there was a light drizzle, I had just switched to high beams...passing the bridge over Ward Creek upstream of Ward Lake. Ahead of me, in the old salmonberry bushes between the forest on the left and the road, I noticed a man-sized, hair-covered figure, light brown all over. For a split-second I thought it might be a deer reaching up on its hind legs to browse, but what I saw next immediately changed my mind.

"I could see the body and head turn to look at me as I drove by less than thirty feet away. The shoulders and head were almost like a man's, but hair-covered. It didn't look right at me and there was no eye-shine. The creature was not especially heavily built and had been walking upright in the bushes when the headlights caught it. I was amazed at seeing one of these creatures you hear about so close to town, but when I turned around at the Signal Creek intersection a moment later and drove back, it was no longer there.

"I waited a good half hour pulled over in my vehicle there, having coffee and watching, but didn't see anything more. I felt safe enough in my truck for that, but I sure didn't get out and go looking for tracks!" [S99B-map5]

The old-growth cedar, spruce and hemlock forest at Ward Lake is certainly popular. It is an ideal location to see deer, river otters, beavers, black bear, ducks, geese and swans, as well as many species of salmon and trout. That in the past forty years there have been at least fifteen sasquatch reports within a mile of Ward Lake does not seem surprising.

The campground one mile north of Ward Lake, Last Chance Campground on upper Ward Creek (a.k.a. Connell Creek), was the scene of one skeptic's startling revelation. A military officer I know, a hale and hearty fellow who until recently would readily laugh at any sasquatch report, passed on to a mutual military acquaintance the following sighting account. As told by the second man, in the summer of 2001:

"He told me he had seen something which changed his mind about sasquatches before moving south, and there was no question about it to him now.... He had gone camping at Last Chance Campground this spring (2001) and told me that a few weeks previously, April sometime, he had been in his trailer late one evening, being the only vehicle there, and got out to stretch his legs and get a bit of fresh air. He was camped close to the creek and it was less than a minute's walk to the highway. There was still some faint light to see by, it wasn't raining, so he hadn't bothered taking his flashlight. He said 'I had just reached the middle of the highway and was looking back down toward Ward Lake and town, enjoying the quiet, when something made me look back toward the forest on the campground side. Just standing there on the near slope of the ditch, about twenty feet away, was a tall, dark figure with broad shoulders. It looked real heavily built, and was a lot taller than a six-and-a-half-foot man, maybe seven feet or more. Right away,' he said, 'it started walking past me swinging its arms, taking real long strides, and just stepped across the highway. It covered the twenty feet in about four steps.' He told me, 'I was so amazed to see the thing, I just stood there with my jaw hanging open until I lost sight of it in the steep ditch on the other side. At that,' he said, 'I just went straight back into the trailer, and didn't hear or see anything more the whole time I was there.'" [S01F-map5]

The friend added that the officer told him that at no time did he smell anything or hear it make any sound. He added the man had the impression that it had been coming out of the thick hemlock beside the highway when it had spotted him, and had just stood there watching him for a moment. Incredibly, it seems that his campsite, or one adjacent it, was the same exact campsite as one other nighttime sighting report around 1982 by a group of fishermen [S82C] and an earlier 1978 report of loud nighttime screaming [V78A].

Near the other end of Revilla's paved highway, Herring Cove, southeast of Ketchikan, has been the focal point of several more recent sightings. Mark S., a twenty-eight-year-old Ketchikan retailer, recently passed on a brief second-hand report from a business friend who wishes to remain anonymous. Mark stated his friend had told him that sometime in late June or early July, 2001, at about 2:30 one morning, he had turned his car around 100 yards upstream from the Herring Cove Bridge. "He told me that there he had seen a brown, hair-covered figure as tall as a man in the meadow across the creek..." Mark stated, "He said it was upright and running across the open area—at least a hundred yards of open grassy flats—'faster than anybody he'd ever seen.' He said it ran into the forest. He seemed serious." [S01E-map5]

Initially, I had not given much weight to this second-hand report, since it had described a figure that could have been any nocturnal athlete in a fur suit, unlikely though that may seem. But the next report, given in June, 2001, shed a different light on the first. Following is a statement from Doug Johnson, a surprised-looking twenty-eight-year-old Ketchikan cannery worker.

"It was the night of June 8, 2001, and I was taking a drive to the (south) end of the road a few miles past Herring Cove to enjoy the clear skies. It was about 9:30 at night, not really dark (Alaskan summer nights are quite short) and I saw this creature crossing the road just before Achilles Creek, from right to left about fifty-five yards ahead, walking on two legs across the gravel road, swinging its arms. It was about six feet tall, I would say, and it was covered all over with dark hair, shiny—like it was dripping wet. It was crossing from the oceanside of the highway, going toward the steep inland side of the road. Before it stepped into the forest I got a good look at how heavy it was built. I'd guess about 300 pounds, anyhow. I didn't slow down and it didn't even seem to turn to look at my car or anything, it was gone in a couple of steps. It was, like, a bit spooky, y'know. I didn't see any other wildlife that trip. One thing I remember was that there was a sort of stink to that stretch of the road. I never stopped later to see tracks or anything, but there is a little clam beach just through the trees at that spot, I've fished there before, a pretty quiet spot." [S01B-map5]

Back at the Herring Cove Bridge, only weeks later, someone else had a very similar experience. Mr. D. H., a taciturn forty-two-year-old Ketchikan road builder, gave me a matter-of-fact interview in October, 2001:

"At about 9:30 p.m. on the night of July 14, I was taking a walk with my wife and son along South Tongass Highway. We had left the house heading north, and had just come into view of Herring Cove on our right. The shoulder of the highway overlooks a steep bank and the curving mud shore of the cove is just ten yards below. We were almost at the bridge when I noticed a dark brown figure below me near the water. It was as tall as a man, running away from me toward the ocean point on the south end of the cove. I don't think my wife or son saw it. It ran on two legs, but it ran faster than a man—faster than a bear in fact. It ran toward the south end of the cove, about a hundred and fifty yards, in less than ten seconds! I saw it the whole time. The tide was out and the sky was clear, visibility was good. It was just starting to get dark and there had been very little traffic. The figure was entirely dark brown and it appeared hair-covered. There wasn't any smell or noise. It disappeared into the bushes and trees near the end of the point. I just kept walking with my family, so I didn't stop to look for tracks, but the tide had come in later that night and by the next morning there was no sign of anything." [S01D-map5]

It would appear possible that the three previous reports are all describing one individual sasquatch, since all three were reported in the same mile of coast, all dark colored, about "man-sized" or six feet. Two of three locations were within 100 yards of each other, and two at precisely 9:30 p.m. These are a lot of coincidences for a hypothetical jogger running that fast in a fur suit in summer. If there

25

was such a person, he should perhaps either be warned of the dangers of wearing such apparel during bear season or be drafted by the U.S. track team.

It is even possible that the first anonymous Herring Creek report may have happened on or around the same night as the road-builder's Herring Cove report of July 14. The fact that there was a sizable return of king (chinook) salmon to the Herring Cove Hatchery in July might be one plausible explanation for a nocturnal sasquatch to follow closely upon the more often-seen evening visitors of this salmon-rich creek—the local black bears.

An immature, six-foot sasquatch, distracted by the noisy abundance of salmon, might perhaps be careless enough to venture sufficiently close to roads and buildings to be observed by passers-by. These three reports may also be related to the mysterious nocturnal lifting and throwing of a 600-pound furnace reported only a quarter of a mile south of Herring Cove in February of the previous year.

Prince of Wales Island Roadside Sightings

The second largest island in Alaska, Prince of Wales Island, has more logging roads than anywhere in Alaska. Small wonder then that it also yields as many roadside sasquatch sightings as its smaller but more populous neighbor to the east, Revilla' Island. In 1999, a Ketchikan hardware clerk, Mr. B. E., told me that he had seen an unusual creature twelve years previously while deer hunting at Labouchere Bay, on the northeast tip of Prince of Wales Island.

"I was hunting opening day, August 1, 1987, at 'Lab' Bay. I had been dropped off up a logging road and had worked my way up beside a clear-cut to a good spot. It was about 11:00 a.m. I was just sitting near the top of the clear-cut, about thirty-five yards from the treeline, when I noticed something black moving on two legs just along the edge of the treeline above me. It was moving smoothly at about ten miles per hour and was just keeping to the trees, moving in and out of them as it went along. I watched it for about ten seconds. It wasn't that big, it was only about five feet tall, but it was upright and moving fairly quick. I knew it wasn't a bear. I saw it move off into the forest away from me. I hadn't seen any deer that morning, nothing came to the call. Right after that, I just decided it would be a real good time to head back on down and get picked up." [S87A-map2]

Only three miles south, around 1984, two other witnesses had described finding two sets of large hominid tracks in snow. Tracks have also been reported in snow nearby, just several miles east of Lab Bay, up Flicker Creek. None of these tracks were small enough to match the smaller stature of the creature reported above, but the notion of a local family group is suggested by the 1984 track report, which mentions two quite different sizes of tracks as "traveling together." (See *Big Footprints on the Alaskan Panhandle*, 1984 by Scott and Bruce Shirley.)

For anyone hoping to see a sasquatch in Alaska from a vehicle at night, it seems the Klawock-Hollis Highway, crossing the middle of Prince of Wales Island, may offer the best bet. It is from the middle stretch between Harris River and Klawock Lake, that two independent reports (Al Jackson in 1996; J. S. in 1999) were forwarded, relating to one 1993 highway incident involving a Mr. I. W. and his wife, then of Klawock, Prince of Wales Island. Although I was not able to interview the witnesses in person, both second-hand accounts agreed in detail, alleging the sighting to have taken place at about 2:30 a.m., while the couple were driving home to Klawock from the ferry that arrives late at Hollis on the east coast of the island. They were driving over the middle of the island, somewhere near the junction of the main highway with Hydaburg Road and the south end of Klawock Lake, coming around a bend in the highway when they picked out a creature crossing the road in front of them. The husband had apparently immediately exclaimed, "Look, a bear!"—to which his wife replied "Yes, but look at that! Since when does a bear walk on his hind legs?" Both accounts agree the witnesses stated it was tall, dark in color and moved quickly. The middle of Prince of Wales Island is quite rugged at this location, densely forested with second-growth hemlock and cedar. This report is just three miles south of another sighting alleged to have taken place two years later [S95A-map4] and one earlier report of an unusual nest. [S93A-map4]

Reports have continued to emanate from this central Prince of Wales Island hot spot almost annually throughout the 1980s and 1990s. In fact they continue today. A brief second-hand report from central Prince of Wales Island came recently through Mr. Dan H., a burly senior forestry professional in Craig, Alaska.

"A local woman, named Michelle, told me briefly that in the fall of 1995 she had accompanied her boyfriend to an area southwest of the south end of Klawock Lake. I believe they went in on the Shaan-Seet logging road via Craig and Port San Nicholas on the west coast of the island. They had gone up the road system about four miles from saltwater. She told me they had seen a sasquatch, dark brown in color, near the head [south end] of the lake." [S95A-map4]

Although details for the previous report are lacking, this sighting was situated fewer than five miles from numerous midisland highway reports, including the following one. In yet another nighttime report near the middle of the Klawock highway, Mr. S. W., an enterprising Klawock businessman, stated the following in a 2000 interview.

"My son and I had been driving the Hollis-Klawock Highway, Prince of Wales Island, late one night during summer, 1997 or 98, driving west, home to Klawock. It was raining lightly and there were no traces of snow in the high places. Somewhere about halfway between Harris Creek and the middle of Klawock Lake, in the area around the south end of Klawock Lake, we spotted a nine- to ten-foot form, dark brown or black in the headlights. It was about 10:00 p.m. and the distance was about 100 to 150 yards. It was moving fast in an upright posture and was crossing the highway to the right, to the uphill side. It went from the middle of the highway to the right, and it sure wasn't a bear or any other animal we have on the island! I looked at my son for just a second and we both said 'Did you see that?' By the time we looked back, it was gone into the forest on the right. The creature covered that distance in only three or four steps! I'll say again, it was no bear. I've hunted a long time in these woods and I've never seen anything like it! Personally, I'd have to say that it was what people around here call a *kushtakaa*." [S97C-map4]

These days in Hydaburg and Craig, on the southern half of Prince of Wales Island, the younger generation usually speak of creatures like the sasquatch as "the hairy man." Stan Edenshaw and his cousin Mickey Calhoun are two such young Ketchikan men, who often travel to Hydaburg to visit their grandmother and family. On one such trip in early July, 1997, Stan and Mickey were halfway down the Hydaburg Highway, passing the trail head to Trocadero Bay around 2:00 a.m. Stan told me the following in a 2001 interview:

"We had just passed the trail sign that marks the trail to Trocadero Bay and were southbound when, forty yards in front of the car, a seven- to eight-foot hairy man stepped into the middle of the road. It never even looked at us. It was chocolate brown colored and covered all over in six-inch-long hair or fur. The arms were swinging fairly straight with only about a twenty-degree angle at the elbow, but it was moving along real fast, left to right. The last two strides it took were about ten feet apart—it covered the twenty feet in two long paces. It didn't even seem to be running that hard, not pumping its arms like a runner would do. It was heavily built, I don't know exactly, but it looked like it would have weighed at least three, maybe four, hundred pounds. We noticed that it hadn't any hair on its face, palms or the bottoms of its feet. Mickey wanted to stop and look for tracks and stuff, but I wanted to keep driving, I didn't feel like stopping—if you know what I mean—and we kept driving on to Hydaburg.

"It was something that kind of shakes you up for a while, so when we got to Hydaburg, we didn't just hit the sack, we stayed up to tell my grandmother about it. She knew all about these hairy men. 'People used to see them regularly around twelve mile in the years past,' she said. We reported it to the police but besides that, only to our family." [S97D-map4]

The same man reported recently seeing a hairy man quite close to the town of Hydaburg. In an interview in 2001, he recalled that in early August, 2000, he had been driving past the gravel quarry one mile out of Hydaburg, late in the day.

"The second one I saw, I was driving past the old Hydaburg quarry last year, which in places is more like a meadow, and looking at it on the left as I drove past, I was a little surprised to see one of the hairy men that a lot of Hydaburg people talk about seeing. It was just getting dark, about 9:30 p.m. It was about fifty yards away, dark brown from head to toe, about seven feet tall, and it was just walking along in the open area, its back to me, away from the highway. It didn't turn around as I slowed for a good look, it just kept walking away, heading west into the trees toward the ocean. I didn't hear anything or smell anything, and I didn't stop, just kept on driving.

"The thing that surprised me was that it was so close to town. All of the old stories, ones I heard when I was younger, were of sightings that people always had of them up the highway a ways, around twelve or thirteen mile, where the road makes a tight turn, there's a cave up there the creatures use there, people in Hydaburg say. But lately a few people have been saying they've seen them closer to town, although not all the old folks take that seriously, but I saw one that night, real close by." [S00C-map4]

Two months later and thirty miles back toward Hollis and Klawock, a Ketchikan woman, who wishes anonymity, reported that in early October, 2000, she and her husband had been driving to the Hollis ferry at about 8:30 p.m. Just as it was getting dark, she said, they saw something unusual cross the highway in front of them. She stated in a later interview: "The creature we saw was large, dark and covered completely with hair. It was walking across the highway halfway between the Hydaburg cutoff and Harris River Campground. It crossed from south to north and was moving quickly on two legs. It was taller than a man, and it wasn't a bear. I know what I saw, and I'll admit I was scared!" [S00D-map4] (see Figure 4)

In 2001 several more reports came to light. In March, there was another report of a sasquatch crossing the Klawock-Hollis Highway on Prince of Wales Island. A respected Gustavus man, formerly of Ketchikan, reported seeing a sasquatch at close range on January 2, 2001. He stated the following in a telephone interview:

"About 5:00 a.m., my wife and I were returning to the ferry at Hollis from visiting friends in Klawock. As we followed the highway through a series of downhill s-turns about one and a half miles from the Hollis ferry terminal, we were amazed to see what appeared to be an eight-foot, hairy, man-shaped creature walking on two legs across the highway. It walked with just a slight up and down movement and had apparently been trying to cross the highway from left to right, but on seeing or hearing us, it seemed to bend down into a crouching position. It was kind of like it was trying to hide from the high-beams. What we noticed was the reddish eye reflection that the headlights picked up. We had slowed down to about twenty miles per hour, because at first we hadn't wanted to hit anything. And looking at it, my wife and I agreed it was no bear. We supposed it would have to have been a *kushtakaa* or sasquatch or whatever you call it. It was interesting all right!"

In a follow-up interview, the witness mentioned that the creature appeared to have one arm up shielding its face, behavior that would be consistent with any nocturnal hominoid facing bright light. [S01A-map4]

In mid-August, 2002, just six miles south of Klawock, Prince of Wales Island, a surprised couple sighted a sasquatch crossing the

Figure 4: Artist's impression of creature sighted by a Ketchikan woman, reported running across Klawock-Hollis Highway, Prince of Wales Island in October, 2000. Sketch by author under witness' direction.

highway, reported Al Jackson of Klawock. Bert C. and his wife were driving north on the highway at about 11:30 p.m. returning to Klawock from a bingo game, when they saw an upright, dark, hair-covered creature walking across the highway, crossing to the inland side of the road in their headlights. They were reported to have stated that it turned as it walked, exposing "long hair running from the back of its head and neck down its back, kind of like a horse's mane." This location is very close to that reported by Stanley Edenshaw, in August, 2002. [S02A-map4]

What these eight reports seem to indicate is that, on Prince of Wales Island at least, sasquatches might be seen almost around the clock, but somewhat more frequently late in the day or at night. Of course, this many reports are not fodder for any self-respecting statistician, and the largest databases, such as those of John Green or the BFRO (Bigfoot Researchers Organization <www.bfro.net>), may suggest something else. The 1987 Labouchere Bay sighting is somewhat unusual in that the reported creature was smaller than most, only five feet tall, and was reportedly seen at midday. The only other sighting of a smaller creature, on Annette Island in the early 1990s [S90B-map5], was also in broad daylight. Two such reports are insufficient to draw any conclusions, but it may be that infant or juvenile sasquatches have different sleep cycles than adult creatures or perhaps have simply not learned to avoid exposing themselves during daylight hours.

2 North Up the Panhandle

Perhaps the first truly documented sasquatch sighting in southeast Alaska dates from May, 1942. John Green noted a report from respected sasquatch investigator Bob Titmus of northern California and later, north coastal B.C., that stated while going to Alaska on a small ship during WW II, Titmus had gone on deck late one evening, probably in Wrangell Narrows, and "saw on the beach close at hand an erect creature about seven feet high covered with dark hair and very heavy." Green also noted, "The ship passed on and he didn't really credit what he had seen."[1] [S42A-map3]

Electrician Paul M. of Ketchikan, described seeing a similar brownish creature in the Wrangell-Petersburg area in the late summer of 1960. He stated in a 1999 interview: "I had been camped out on a fishing trip near the mouth of Ohmer Creek, eighteen miles south of Petersburg by road, near Blind Slough, and had been near my tent by the creek after dark, when I heard splashing just a few yards up Ohmer Creek. I shone my flashlight level, and very close to me, just a few yards away, were two long, hair-covered legs, brownish in color. They were a lot longer than a bear's. I could just make out a large body above, but didn't shine the light up in the direction of the head. Instead, I just decided to turn and head for my car! I didn't see it again after that." [S60B-map2]

Mr. Steve M. of Petersburg, Alaska, in a 1999 interview, reported a group sighting around 1968 on Wrangell Narrows between Kupreanof and Mitkof Islands, just south of Petersburg.

"It was November 1968 or '69 I believe, an overcast night, and I was boat camping with four friends. We were all in junior high and had camped out on a point on Kupreanof Island directly opposite Woody Island. We had taken guns, had a shotgun and had been target practicing all day and into the evening. We had our tents set up onshore near the boat and were just building up our campfire, when we heard crashing in the bush—the sounds of trees being broken. Some of the guys thought they saw a large, brown, upright figure and then right away we all heard a strange sound. It was a real loud thump, like a medium-sized log being swung against a big log or tree. That was followed by another loud thump, and we bailed out of there! We just grabbed our tents and gear and jumped into the boat, heading across the couple hundred yards to Woody Island. We figured we would be safer and camped there. Nobody saw any other people or heard any shots or human sounds. Some guys said it had been a bear but others of us disagreed, saying it was a *kushtakaa*." [S68A-map2]

The reported sighting was brief, but the wooden "rapping" sound is something mentioned in other sasquatch reports and will be touched on later.

Another more-detailed report from the Wrangell region surfaced in about the same year. In George Haas' *Bigfoot Bulletin*, no. 11, Nov. 30, 1969, is an interesting letter from Mr. J. W. Huff, of Ward Cove, Revilla' Island, Alaska, containing the following excerpt regarding a more peaceful sasquatch sighting above Bradfield Canal, north of Wrangell, in July of 1969.

"I have been working for another mining company during this past season and one of my camps was in the Bradfield Canal general area. To be more specific it is located on Bradfield Canal Quad map [B-5], T-64 S, R 88 E, Sec 23, N.E. Cor. at elevation 2,300'. We were flown in by helicopter late in the afternoon and as soon as the 'copter was unloaded it immediately departed for Ketchikan. We started to get our camp up and prepare for the coming operations. While working on the camp shortly after the departure of the 'copter I happened to look up on a ridge about 300 feet higher than we were and about 500 yards distant. I saw a man standing there watching us and remarked to my partner that I thought we were in someone else's area and would have to move out the next day. My partner and myself watched this man for ten or fifteen minutes expecting him to come down to our camp and visit with us and inform us this was his area and, of course, we were prepared to tell him we would leave the next day. He stood absolutely motionless and we had a good look at him and he appeared to be extremely large.

"As we kept watching him he seemed to have no hat of any kind and looked very dark and we had the impression he had no clothing on. After watching him, I resumed my work in getting the tent up while my partner continued to watch. He then yelled to this man and waved to him whereupon this man took off with a lumbering gait rather rapidly and soon disappeared over the ridge from our sight.

"We looked several more times expecting him to appear somewhere else coming down to our camp but no further sighting was made. We did not have our radio antenna up yet, or I would have recalled the 'copter to investigate further.

"The next morning we had to decide where we were going to start our prospecting so we decided we might as well start where we had seen this huge man. It took us some time to get there as we had to take a roundabout route and were prospecting all the way. It was very warm and rainy so the rocks were pretty slick and we were taking our time. When we got to the place where this man had been standing, there were no tracks, of course, as it was bare rock. Nearby was a snowfield and we searched this area but could not find anything we could definitely pin down as tracks. There had been goats up on that same ridge the day before but we could not find their tracks either due to the melting condition of the snow.

"We did find some depressions in the snow that would require quite a bit of imagination to definitely call them tracks. This snowfield was rather small and once across it bare rocks were once again encountered. What this thing was that we saw I do not know except that it was no known animal that I have ever encountered and all I can say is that I have never seen anything even approaching this thing. We were in that camp for about three weeks and never again saw a trace of him.

"I am writing this letter in answer to your request for information and have related these experiences to you just as I saw them and whether or not they are believed I personally could care less as I know what I saw or at least what I appeared to see. I am fully positive that there is no reason that these sasquatch could not exist as there have been too many reported sightings of similar man-like creatures." [S69A-map3]

This report has been noted previously by other researchers, and although it is typical of the curious behavior attributed to sasquatches in some reports, it is rather remarkable for the length of time the witnesses were observed. Perhaps the creature had never seen a helicopter land before. Perhaps the tasks of the men were of interest. At any rate, this reported "watching" behavior is not usually observed, since in most reported sightings the creature apparently has some other place to be, and perhaps, as in our species, some individuals are just not as curious as others. Prolonged observation of people by a sasquatch is also noted in a policeman's report from Yakutat, Alaska. (See *Close Encounters*, sighting by Fred Bradshaw.)[S70A-map1]

There are several reports of an alleged sasquatch (or sasquatches) exhibiting aggressive, or at least curious, behavior toward humans. The behavior of the tall creature reported by Mr. Huff might easily be described as depicting curiosity. The significance of the wooden log or trunk "rapping" sounds attributed to a sasquatch by some of the Steve M. party near Wrangell is unknown, but may have represented locator behavior, curiosity or even aggression.

In March, 2000, a staff member at the University of Alaska, Ketchikan, mentioned to me that a friend of hers, Mr. H. K., living in Kake, a small town west of Petersburg, had reported seeing a sasquatch on the outskirts of Kake only weeks previously. The man in question politely responded to a request for an interview and by telephone related his recent experience.

"It was on February 16, 2000, about 10:30 p.m., and my girlfriend and I were driving north into town on Keku Road, just near the RCA building. We were driving in snow up the hill when, twenty-five yards ahead of us, a dark, hair-covered figure sprang up from the ditch. It was about five feet tall, but was crouched over as it jumped up the bank out of the ditch. It covered a distance of about six to eight feet in one bound. It didn't even look at us, which I thought peculiar, and there was no eyeshine like an animal might have. It swung its arms as it moved and appeared to be black or dark brown all over. It was thickset and had wide shoulders; we're sure it was no bear! There have also been other reports around town at this time of the year over the past three years.

"There were also some real low tides about then and there are lots of clamming beaches, some of the best are right in town. The fellow who does roadside cutting and cleanup for the town, Mr. R. J., told me he found some large tracks in snow there the next day, just a hundred yards up the road." [S00A-map2]

The height given in this report would seem to match the description of an adolescent or older juvenile sasquatch but the "hidden height" factor in a crouched posture may have placed the actual height at six feet or greater.

There are four other accounts of hair-covered hominids further to the north in the panhandle, three of them detailed in the following chapter, "Close Encounters," and one from Prince William Sound, which lies beyond the north end of the panhandle. Although that is just beyond the northern geographic scope of this book, it is included here. Dr. Henner Fahrenbach[2] forwarded to me, through Ray Crowe, the following report given him from a salmon fisherman in 2000.

"Ed L., a commercial fisherman, stated that in early fall, early 1990s, he was salmon fishing with a companion in Prince William Sound. After anchoring offshore, his companion took a small boat up a river to check on the state of the salmon run. As the day wore on toward evening and he didn't come back at the expected time, Ed scanned upriver and across the adjacent land with binoculars. There he saw a sasquatch walking across the tundra with long, smooth steps and with dark hair flowing from his shoulders, bouncing behind "like a cape" at every step. The sasquatch paid no attention to the boat [distance about 1,000 feet]. The area, according to Ed, was known for its concentration of berries and black bears. The salmon run was in its early stage. Solid forest was estimated to be about 100 miles away. Ed did not have even a shadow of a doubt about the identity of the animal he had seen, particularly since he had previously hunted many of the large game animals of Alaska. Hair of the described appearance can be estimated to be eighteen to twenty-four inches long." [S90A-map1]

While it may appear from the relative scarcity of reports from the northern panhandle that there are fewer sasquatches to be seen there, this is probably misleading. In "southeast," as the panhandle is called, reports are hardly ever given to media journalists and usually propagate only within a word of mouth radius of one's immediate family or, at most, hunting or fishing partners. Everywhere there is a tight-lipped inclination to be wary of the adverse effects of ridicule, much more damaging in the smaller villages and towns of rural southeast. I don't doubt that were one to spend a decade or so in each of the panhandle's smaller northern communities, far more reports would come to light. As it is, Ketchikan and Prince of Wales appear to have a much greater population of sasquatches than areas northward, but this could be false conjecture. The panhandle's climates, vegetation, salmon and shellfish populations are all fairly uniform with what hunters and fishermen get elsewhere in Alaska. More research and documentation is definitely indicated throughout southeast Alaska to obtain even a basic estimate of sasquatch populations there. In the meantime, a closer look at more detailed reports will perhaps eventually be useful in helping biologists and anthropologists plan their research. Finding an object in the dark is always easier if we are familiar with details of that object first-hand at close range. To this end, let us take a few cautious steps closer, for some more glimpses of the sasquatch of southeast Alaska.

1. Green, 1973, *The Sasquatch File*, p. 17.
2. Fahrenbach, H. F., in *The Track Record,* Journal of the Western Bigfoot Society, Hillsboro, Oregon.

3 Close Encounters

Encounters on Revilla' Island

It is evident that sightings of creatures by people in vehicles, walking along or simply sitting still outside, yield progressively greater amounts of detail and information. In a January, 2000, interview, Ms. Shannon H., a Ketchikan woman, described what seems to have been a close encounter of the uncomfortable kind. It was something, she said, that she had not told anyone for a long, long time.

"On a weekend in August, 1966, I had gone on a camping trip at Ward Lake, Revilla' Island, with two girlfriends and our five young children. It was a clear evening and we were camped on the lakeshore part of the Signal Creek Campground right beside the creek culvert. It was around ten or eleven in the evening and was getting dark. Our kids were all asleep in the tents beside us and we were relaxing together at the picnic table beside our fire. I was facing the lake, but T. was facing the creek and K. was dozing when T. exclaimed, 'Look, up by the creek, a bear standing up!' We were pretty much alone, the only other two campers were on the other end of the loop toward the forest, and we were alarmed, so we all grabbed our flashlights and stood up by the car and tents and shone our flashlights up the creek. Sure enough, about forty feet away in the alders beside the creek, stood a tall, hair-covered figure. But instead of the bear we expected, we were surprised to see a seven-foot-tall manlike creature with long legs, really heavily built, and covered with black hair.

"As we turned our lights on it, it turned first its head, then its shoulders to the left, as though it were turning its eyes away from the lights. The shoulders were wide, the head seemed to sit right on the shoulders and it had long dirty-looking dark brown matted hair hanging down on either side of its head to the shoulders, almost like dreadlocks. Its face was flat, not like a bear's. It started angling toward us with a shuffling kind of walk. At that point we hit the panic button, threw all the kids in the car, stuffed the tents in the back without taking them down and drove out and around the lake to spend the night in one of the three-sided picnic shelters with fireplace and tables.

"By eleven-thirty or so we were settled in and had all the kids asleep in a corner of the shelter. The three of us got into our bags around them and were all asleep by midnight.

"Sometime after that, T. and I were awakened by the noise of heavy footsteps approaching the shelter. There was heavy breathing, and T. heard it too. We were petrified. We could see the creature in the entrance, smell it and hear its breathing and footsteps as it walked right into the shelter. K. was still sleeping, but T. and I could see each other's faces as we lay in our bags between the creature and our kids. It had a real musky smell, like a musk and earth mixed together. It was tall and dark, it walked around in there for about four minutes, then we heard it pick up our ax.

"I can tell you, T. and I were scared! It seemed to be playing with the ax, but I don't know if it knew what it was for. It never chopped anything with it, or swung it at anything. It just seemed to want to drag it around on the floor. It never made any sound besides a kind of slow, huffing breath sound, with the ax scraping.

"T. was just huddled in her bag like I was, hardly daring to breathe. After about five minutes, it walked out of the shelter and didn't come back again. I don't remember sleeping again that night, we didn't say a word that night, and in the morning T. and I said about three words to each other, that was it, and we packed everyone out of there.

"The ax was gone. On the beach on the way back to the car we found wide tracks about fifteen inches long but we didn't stop to look at them. While we were leaving, two park rangers came around to ticket us on how we were camping illegally in the shelter, so we told them about the creature we had seen at Signal Creek Campground and how the thing had followed us over to the picnic shelters. They just looked at us for a moment, amazed like, and told us we had 'just seen a bear.'

"Whatever it was walked on two legs, swung its arms when it walked, and I know a bear when I see one! T. and our families still go out picnicking, but she doesn't bring up the subject and, for that matter, I don't either. People just don't want to hear about it, they ask you what kind of drugs you're on!" [S66A-map5]

When interviewed, the woman gave the impression of being sincere and relieved to have at last found someone who would take her story seriously. At the same time she seemed hesitant to call up memories that were fearful to her, although nowhere in her story did the creature actually behave in a threatening manner. She didn't seem like the sort who would panic easily and on re-interview a week later, recounted the details exactly as she had the first time. When asked why they had not actually left for town after the first sighting, she replied that the building seemed quite safe, being much closer to the main road to town, and that the children would be a lot easier to get back to sleep if they didn't have so much time to wake up. She said they never imagined it would follow them around the lake.

Coincidentally, the other witness, Ms. T. W., was known to me. She was initially reluctant to relive the event in an interview, but eventually did so and confirmed the whole account in detail. It does seem unusual that the reported sasquatch followed them the half-mile around the lake. Curiosity is certainly an appropriate term for the alleged manner regarding the missing ax, behavior which fortunately ended without harm to the campers. But perhaps this behavior represents something else more romantic. Its activities in the shelter with the ax, which was found to be missing in the morning, might even be called "play."

Less than a mile upstream from Ward Lake is Last Chance U.S. Forestry Service Campground, the location of numerous other sasquatch reports. Mr. Steve S., a lanky Ketchikan commercial fisherman, recounted a hair-raising encounter he had twenty years ago with several teenaged friends. Steve agreed to an interview in January, 2000, his eyebrows arching as he recalled the events some twenty years previously.

"In May, 1982 or '83, I was camping at Last Chance Campground on Connell Creek for steelhead with [friends]. The very first night there, we all awoke around midnight to a splashing in the creek. Thinking it was maybe a run of fish or a bear, we all got up with flashlights to take a look. Our tent and a motorcycle were only

Figures 5 and 6: Photographs of Larry Thomas at the scene of a 1992 reported sasquatch sighting near Ward Lake, Revillagigedo Island, AK.

a few yards from the creek; our vehicle was parked in the campsite beside the tent. In the moonlight less than twenty yards away we saw a seven- to eight-foot form shaped roughly like a man, but with long legs and a really heavy build, standing there facing us. I remember we were in such a hurry to get out of there that we ran right over our motorcycle and knocked it into the creek! We left in a hurry and didn't go back to get it out and pack our tent until the next day." [S82C-map5]

In 1992, Revilla' Island's sighting hot spot between Ward Lake and Connell Lake was again the scene of another sasquatch report. In an interview in 1996, a retired Ketchikan equipment operator, Larry Thomas, spoke of a startling encounter that caused him to draw a gun.

"In July of 1992, I was doing security work on the Pipeline Road, now Revilla Road just north of Ward Lake, eight miles north of Ketchikan. I was working the night shift driving the [then] gravel pipeline road that goes some dozen miles north from the mill. The job included driving and checking all the locked pipeline access gates. It was a clear night, I had just driven up to gate two and the time was approximately 2:30 a.m.

"Nearing the gate, the vehicle headlights picked up something that made me stop to investigate. About seven feet off the ground was a pair of eyes that were shining whitish, not amber, red or green, right beside a medium-sized hemlock. To see more clearly, I slowly drove forward until I was only about fifty feet away, and keeping the eyes in sight I slowly opened the car door and eased out. With the parked headlights and my flashlight, I could make out a dark upright form with head, arms and upper body visible. It was still standing and waving its arms to pull the bark off the hemlock. It did not appear to be a bear. All the while the eyes were still glowing in the headlights, they were glowing whitish and appeared completely round.

"I could make out that it was chewing or mouthing something. At that point it ducked down from the place where the bark was torn off and started walking, manlike. It was walking to the left, so I whistled loudly and it startled and started walking on two legs right toward me on the road. I noticed it swayed slightly from side to side as it walked. At this point I was carrying only the flashlight and a twenty-two- caliber handgun. When I stopped whistling and backed up toward the car, which was still running, it turned to the northwest side of the road and into the bush. At no time did it drop down on all fours.

"I reached into the car, got on the radio and told the gate guard who simply advised me to get back in the car and continue. Several days after that, another security guard, who was a roaming guard for the same company, had a word with me. When I told him what I had experienced, and that for whatever reason the management was officially passing it all off as a bear, he said to me, 'That was no bear, Larry, that was a *kushtakaa*.' The other guard [name withheld] told me 'I don't care what some say, a while back I went through the same

Figure 7: Author's impression of sasquatch reported drinking at a stream near Saxman, Revilla' Island, AK, in 1978.

gate, gate two, got out to lock it and heard a noise. Something hit me on the side of the head and knocked me out. When I came to, I found a rock beside the vehicle. Another night soon after I saw it feeding on grubs from a log below the back of a building. I reported it, and they told me not to repeat it; they said it was just a person.'

"The other guard told me that he had driven out to the spot during daylight soon after and found tracks and an area where grass had been pushed down. The other guard said his dog had snarled and put his ears back, facing toward the river [Ward Creek]. There were several large branches lying around that had recently been broken off a nearby tree at a height of about six feet off the ground.

"Something else about that night in July, '92. Just five minutes or so before seeing the creature near gate two, I had seen some big animal crossing the road a few hundred yards ahead, but I just dismissed it. One of the foremen told me later, 'You want to be careful, I guess you know about this *kushtakaa*!'" [S92A-map5] (Photographs of the report location appear in Figures 5 and 6.)

Saxman, five miles south of Ketchikan, was the location of another report of a close encounter. George J., of Ketchikan, shared the following second-hand account in 1993.

"In the fall of 1978, when I was twelve, at the start of football season and when the pinks [salmon] were still in the creek, my friend Carl S. told me he had been up the second creek south of Saxman. He told me he had been cutting through the forest to the creek and had been coming up over the bank to where you look down on the creek. He said he saw something. 'It was a big, dark, furry creature just yards below' him and it was drinking down at the creek, all bent down. 'But it was drinking by cupping its hand and drinking from it!' He added he had run home and told his grandma and ma who immediately told him 'not to tell a word to anyone, not then or ever.' But he was my school age chum and he did share with me." [S78A-map5]

The boy has since grown and left the area, but in an interview given by his mother in 2000, she corroborated his story.

"He certainly saw something unusual, and his is not the only report from around here. I definitely believed him. It was what he said he observed about the creature drinking from a cupped hand that was especially convincing."

If heading home after an encounter was all a witness did, then most reports would end peacefully enough. But many Alaskans carry rifles or guns. In southeast Alaska, there have been several cases where hunters have apparently lined up their sights on, or even shot at, a "bear" that wasn't. One such situation occurred with Baranof Island fisherman and hunter, Mr. H. A. (whose alleged close encounter with a sasquatch on his boat is noted further on). He mentioned in a 1999 interview that while deer hunting with companions in the fall of 1997, on the Baranof Island side of Walker Channel, situated approximately forty miles south of Sitka, they had run into what they had at first thought was a bear. He recalled the surprising encounter for me.

"We had just come up over a rise to this open area in the forest and saw this big, upright figure at about forty yards away. It was tall, about nine feet, and covered in dark brown fur. We changed our minds right away though, and lowered our rifles, because as we watched it watching us it became clear that it was no bear. At first it had just stood and watched us, but after about one minute, it walked on two legs taking good strides into the bush." [S97A-map2]

Retired Washington law enforcement officer Fred Bradshaw wrote me in 1999 of a series of more upsetting sasquatch encounters he had been informed of while living in Yakutat, northern Southeast Alaska, in the late 1970s.

"In or about 1979, I was told [of an encounter] by Yakutat Native Steve Johnson and others who picked berries at the end of Dangerous River, which runs thirty miles northeast of Yakutat past the airport. There is a lake there called Forrest Lake. On the backside of Forrest Lake a sasquatch was living, and when they hiked into Forrest Lake to pick berries this sasquatch would show up and watch them. It was seven feet tall and appeared to weigh 600 to 900 pounds. It would watch them and kind of stalk them, and the Natives, scared, would leave. A bunch of Natives went up there and shot at it with shotguns and ran it off. But it came back, so the Natives wouldn't go back there. The last I heard the creature is still there. It has scars on its back from the shotgun blasts, but it doesn't bother anyone." [S70A-map1]

I haven't flown to Yakutat, so I cannot confirm or discard the policeman's story. The account is rather amazing and at first glance has more than several elements of good storytelling. Before passing it off as a good yarn, it may be worth noting that there are no more reports of sasquatches falling to a bullet than there are of sasquatches being apparently hit, then walking or running off. It would appear that a certain proportion of sasquatches alleged to have been shot might be able to make it away without eventually dying of the wound. We humans are much smaller than reported sasquatches, but there are numerous individuals walking the streets after each war with fairly good-sized chunks of lead in them. The point is, any animal weighing 600 to 900 pounds is not necessarily going to die from being shot, nor is it necessarily going to leave its home range. With the glaciers on one side of Yakutat, and the ocean on the other, there just aren't a lot of other places you can go to. If berry-pickers thought Forrest Lake was worth going back to, perhaps a sasquatch might as well. As to "stalking" or possible curiosity behavior toward humans, the extreme scarcity of human beings along the edge of the St. Elias Range might account for that—so might a local scarcity of other sasquatches, for that matter.

On rare occasions, some hunters have stated that they shot at what they at first thought was a bear. In a startling 1999 interview,

Ketchikan hunter Steve B. confided that he might possibly have wounded a sasquatch on the north end of Minerva Mountain, two miles north of Ketchikan (Bear Mountain) in August of 1996.

"On that day, I was close to the top of Bear Mountain on an open area near the north end of the ridge that runs north toward Ketchikan Lakes. I was scouting for deer, but was actually geared for anything, as I was carrying a .300 Weatherby magnum—a good, flat-shooting rifle also handy enough for bear—for which I had a license as well. It wasn't a bad day at all, no rain, and I had just been sitting down for a break and to watch for game.

"A curve of black fur showing over the top of a big log about one hundred yards away caught my attention. Thinking I had a good spine shot on a black bear that was rooting for grubs or such behind the log, I took a shot at where I could hit the backbone, not waiting to change my scope from two power.

"Immediately on firing, I saw the animal stand up and without stopping, walk quickly on two legs out of the open area into the timber. I thought I must have missed, or at most just pushed the bullet through a nonvital area, because the animal showed no sign of having taken an impact. It didn't knock down or anything. It just got straight up and walked away, kind of like a man, toward the closest trees. I have never seen any bear do that.

"I was so amazed. I dropped my rifle on the ground! It was pretty hair-raising. I just stopped a while, relieved that I hadn't hit it. A while later, after I had collected myself and my gear, I looked over the area but couldn't find any blood or sign. It had looked pretty heavy built, no neck, black fur all over, about seven feet tall and had swung its arms when it had walked into the trees. I never smelled any odor and I never heard it make a sound. I'm pretty sure I didn't hit it, and that's okay with me because I don't think I'd feel too good about it if I had." [S97A-map5]

It is somewhat remarkable that, in spite of periodic doses of flying lead, sasquatches do not seem to be very good at measuring up to their ferocious tabloid image. Some witty pundits have enjoyed pointing out that there don't seem to be any people claiming to have been injured by angry elephants either, since any such hunter returns victoriously unscathed or becomes part of the landscape. This line of thought, while not without the occasionally rumored story to lend the idea spice, seems to be quite unsubstantiated. It appears that sasquatches are as difficult to anger as the gentlest of the great apes, the gorilla, and just as in that species, sasquatches indulge in threat displays simply to bluff any animal threatening them, thereby avoiding conflict. In most reported encounters, the sasquatch does not usually allow the witness to approach for a closer look.

But, in 1998 James Edenshaw of Ketchikan, a forty-four-year-old Tlingit and Haida artist, told of a close encounter he had with a large, hair-covered creature in 1982 and illustrated the face in an accompanying sketch. The setting was less than one mile east of the previous Bear Mountain report, in the hills just north of Ketchikan's Bear Valley suburb.

"It was the fall of 1982, and I had left the house early one day to explore the forests and caves north of Bear Valley. I had left early and headed north over the ridges that separate Bear Valley from Ketchikan Lakes. After a couple of hours, I had come onto a sort of knobby hill, pretty well covered in trees. In the rocky side of the hill I noticed a passageway about thirty feet long, sloping down with daylight showing faintly at the other end. I was curious and made my way into it. It was just big enough for me to crouch along and sloped gently down to where I could see some light at the other end. Approaching the end of the passageway, I could see it had a sort of small opening that had a view into a cave about twenty feet wide. I could see a way through the larger far cave opening to the forest on the other side. But what I saw partly concealed in the cave just on the other side of the small opening made me stop.

"About eight feet from me were the face, shoulders and chest of a creature, eyes closed, sleeping or resting. It was sitting, its back to a rock wall. I could see that the floor of the cave was about five feet below me, and as I watched the creature, it looked big. Its head was about even with mine and the top of its head must have been about five feet from the floor of the cave. Its shoulders looked about four feet across and covered with long dark brown or black hair. I could see its face from the forehead down, its chin resting on its chest.

"I watched it, real quiet. The face was clearly visible from the big ridge above its eyes which were closed. It didn't have much hair on the face; you could see the skin, kind of leathery and dark gray. The nose wasn't quite flat. It was broad and apelike but it was actually kind of convex and flared into wide nostrils. The jaws were big but didn't stick out that much. He was resting or sleeping. You could

Figure 8: Witness James Edenshaw's sketch of face of large male sasquatch reported seen sleeping in a cave near Lower Ketchikan Lake, Revilla' Island, AK, in 1982.

tell it wasn't a big human or a bear. It seemed to have its legs stretched out in front of it but I couldn't tell exactly. The lower edge of my opening blocked the view of his chest and lower half. He wasn't making any noise and neither was I.

"His hair was about four to six inches long, maybe longer on the side of his head. I couldn't see any ears under the hair and I couldn't see the back of his head or the back of his neck. In fact, I couldn't make out any neck at all, just huge shoulders and upper arms. There was a big patch of hair on the sides of his cheek and jaw which was grizzled white or light gray. In the cave I thought I saw a couple of old wrecked things that looked like old pieces of mining equipment but I couldn't tell for sure, it was mostly just rock and gravel.

"After looking at him for about ten seconds or so I reasoned that it might be a good idea for me to get out of there. Although the small opening I was looking through would have been way too small for him to get through, I just backed up the way I had come in, just moving real quiet and careful so as not to make any sound. It took me about thirty seconds. When I got back out, I just headed back the way I had come and went straight home." [S82B-map5]

Mr. Edenshaw's report concurs with descriptions generally

Figure 9: Sasquatch reported climbing on board a boat off Prince of Wales Island, AK, around 1951. Illustration by author, under witness' direction.

given for large male sasquatches. The only really unusual point mentioned here is the situation of the animal in a cave. While Alaska does not generally produce such reports and the larger database does not appear to indicate caves as typical shelter for sasquatches, there is one report of an unusual nest found in an abandoned mine shaft ten miles east of the above report. If sasquatches were the opportunists they appear to be however, there is no reason they might not choose to utilize such shelters for brief periods, especially in an area where the annual rainfall can exceed 150 inches.

Prince of Wales Island Close Encounters

There are at least three reports of sasquatches climbing onto boats at night in southeast Alaska, and the following is one such story. It was related in 2000 by Mr. R. B., a fit-looking fifty-seven-year-old Ketchikan demolitions professional. He was involved in a sighting in 1951 or 1952 when he was a boy, "somewhere on Prince of Wales Island."

"I was working with my grandfather on our thirty-six-foot fishing boat, the *Verna May*, somewhere on Prince of Wales Island. I don't remember the exact location, but I do remember us anchoring up close to shore one nice evening and going to bed. It was after midnight, and I had got up to go on deck, as we had no facilities inside in the cabin. As I flipped open the cabin door onto the deck, I felt the boat rock, and right in front of me, in my flashlight beam, was this tall, hair-covered, manlike creature, climbing up over the side of the boat, all wet. He was dark brown or black. As soon as he saw me, he just kicked off backwards into the water with a big splash. Granddad came up quick and looked with me, but it wasn't swimming on the surface anywhere. Then, just a few seconds later, we could see this seven- or eight-foot figure stand up out of the shallows about forty yards away. It stood up on two legs and walked straight up into the forest. It sure covered that distance underwater fast! Granddad saw it stand up too, and he got real scared. He pulled the anchor and started the engine right away and we got out of there. We kept moving for a time until we got to another anchorage late that night. In almost fifty years on the water and in the bush that was the only time I seen one of those things!"

This is one of at least two accounts from the panhandle where a sasquatch was reported climbing on a fishing vessel at night, startling the skipper into relocating his anchorage. What might prompt such behavior, besides curiosity is not known, but theft of fish is commonly found in both Native folk tales and modern accounts. The rocking of boats is noted in several other similar reports, and the creature's quick disappearance on being noticed seems typical. Its style of swimming is commonly noted as submerged, not the on-the-surface style one might expect to hear for any ape or other primate, or the crawl style if one were to think of a human. (This ability to swim powerfully underwater is also noted in old Haida and Tlingit stories.) In these and following reports of sasquatches in water, there is no mention of the use of arms in swimming. Several anthropologists examining purported sasquatch tracks in mud have pointed out some indication of partial webbing between the toes, and if that is the case, then the sasquatch already has "swim fins." As such, no arm motion would be required for strong propulsion on or under the water. This may be significant, as there are reports of sasquatches "popping up" near boats or islands considerably farther offshore.

Close Encounters "North up the Panhandle"

At a September, 2001, organizational meeting in Ketchikan for a prospective documentary film dealing with *kushtakaas* and sasquatches, a young woman spoke out. She reported that following a hike just west of Juneau in early July, 2001, she and two friends had found themselves face to face with a startling, upright creature at the edge of their bonfire. In a steady voice, she related the following.

"A Juneau friend had just taken us on an evening hike up a trail off the end of a gravel road west of town and we had come back down. We were resting at a campfire we had near the foot of the trail, not far from our parked car. It was about 9:30 or 10:00 p.m. and was getting dark, when we heard the scuffling sound of footsteps coming out of the forest about twenty-five feet away. In the light of the fire, we looked up to see a black, hair-covered figure, about six and a half feet tall, with very broad shoulders. It had one arm kind of held up over its face as if to shield its eyes, and was edging close to us—peering at us. We called out, 'Who are you, what do you want?!' But it didn't answer, and we began to get real nervous as it edged closer.

"It was covered all over with black hair, and the eyes reflected the firelight with a whitish, then as he moved by the fire and we started to move away, a reddish color. We were standing about fifteen feet away. I couldn't see his teeth, his mouth wasn't open and the whole time, he made no sound at all. His body was more ape-like—hair all over—but his face was long and narrow, with ash gray colored skin showing. We were still shouting, 'Who the heck are you?' and starting to back up. I was getting hysterical, and we all broke and ran for the car, yelling. Running past the back of the car to get in, I noticed a huge hand print in the dust on the back window. We didn't look back, and just pulled out and headed down the road back to town.

"I remember it had a bad smell to it, like wet dog and garbage mixed. It was heavily built, but not threatening really. The total time he looked at us was about ten to fifteen seconds, I think. It scared us all." [S01C-map2]

When I showed the woman the grouping of four illustrations of different fossil hominids and *Gigantopithecus* that evening, asking

Figure 10: Facial appearance of sasquatch reported approaching three people at campfire north of Juneau, AK, 2001, illustration by author.

her to select the individual facial anatomy that most closely resembled the creature she had seen, the woman immediately indicated *Zinjanthropus*, a large, robust type of Australopithecus, known from man-sized African fossils.

On a second interview a week later, she repeated her story, adding only a few details. Again she confirmed the facial features of *Zinjanthropus* as the most similar of the four hominoid reconstructions. Her particular selection should not be that important, primarily because there is no one commonly accepted reconstruction for any fossil hominoid, at least when one is forced to speculate on skin color and hair distribution, length and color. It is interesting to speculate however, if there may be more than just one type of hominoid responsible for the bulk of reports in "southeast."

Undoubtedly, the closest alleged encounter with a sasquatch in southeast Alaska is that of Mr. Harold A., a Sitka commercial fisherman. Following is his 1999 account of what happened in the summer of 1979 when he was working on the west coast of Baranof Island, about thirty-two miles south of Sitka.

"We had our fifty-two-foot seiner, then getting some use as a fish packer, up at the south end of Cedar Pass, and it was late. We were just finishing anchoring up for the night about 100 yards from shore and we were settled. My crew was below tending to securing gear. I was alone in the cabin, resting at the galley table with the starboard porthole window open above my elbow, with the cabin light on as usual. It was dark outside.

"Just then, I felt something lift my right arm and looked up to see a huge hairy arm, bigger than a gorilla's, reaching through the porthole and lifting my arm. Its arm was covered with dark brown fur and I could see the whole forearm up to near the elbow. The hand was palm up and huge, being about three times the size of a human hand. The skin was a dirty brownish gray color. The whole of the arm except the palm side of the hand was covered with four-inch, dark brown or black hair. I could not tell if there was a thumb. I wasn't being squeezed or anything, just lifted up, and I would have to say it felt gentle. But I can tell you I hollered and pulled my arm away, and I didn't even look outside! I got the engine started and the crew to pull anchor and we moved north up the channel about five or six miles to a new spot before deciding that would be a better spot to anchor for the night. We didn't hear or see anything else that time." [B79A-map2]

The report was later corroborated by a nephew of the witness, and did not differ in any manner from the eyewitness account. The captain was calm-mannered during the interview, but still apparently a little amazed by it all. He did not hesitate to emphasize that his initial wonder was quickly replaced by alarm and an unexpected departure to a whole new anchorage. He admitted that the crew down below had been surprised when he ordered them to pull out again, as it was dark and they had just finished the task of setting anchor in what, on the charts, appeared to be a very nice, sheltered spot.

His descriptions of the large hand size and grayish pigmented skin accord well with other close-range observations of sasquatches. Two things in the report are striking as supportive of typical

Figure 11: Fishing skipper Harold A.'s close encounter with a sasquatch on board his boat at Cedar Pass, south of Sitka, Baranof Island, Alaska, 1979. Sketch by author from description by witness.

sasquatch anatomy and behavior. One is that, although the skipper states his arm was picked up by a large hand, its thumb was not in any way noticeable to him. While this might not seem likely were a human to lift another's arm, the sasquatch thumb has been mentioned as being set further back, not as fully opposed as a human's thumb. This attribute has been studied in detail by Krantz[1], working from a collection of casts of sasquatch hand imprints. It is his conclusion that the thumb may be fully opposable, but not with sufficient palm musculature to rotate with great strength. The function of a sasquatch grasp would not be compromised by this in any way, as there are a number of other grasps that all primates use with strength and precision.

The other interesting attribute mentioned is the gentle nature of the animal's actions. The restraint alleged seems to accord closely with the noticeable absence of aggressive behavior reported elsewhere when a sasquatch is not threatened or approached suddenly in what is hypothetically its "core home range." The skipper's story would appear to be just another case where a sasquatch exhibited the same sort of curiosity we ourselves have toward members of a similar species.

The number of nighttime observations brings up the thought that if a creature is adapted to, or at least used to, nocturnal activity patterns, bright light is something a sasquatch would avoid. Two reports from southeast Alaska mention a sasquatch lifting its forearm over its eyes as if to shield them from the glare, in one case from firelight, in the other from headlights. It is also noted frequently throughout the reports how witnesses were surprised at how the creature seemed to deliberately avoid turning to look at the oncoming vehicle. Where sasquatches were said to face the light, eyeshine similar to camera "red-eye" was noted. Hypothetically, this may be related to a richer bed of capillaries in the sasquatch retina, allowing improved nocturnal vision. It may be difficult to prove without a body. The question is largely anatomical, and I am not a sufficiently large anatomist to answer it.

1. Krantz, Grover, 1999, *Bigfoot Sasquatch Evidence*, Hancock House Publishers, Surrey, B.C., p. 154.

4 Aquatic Anthropoids

For the anthropologists familiar with Alaskan Native traditions of hair-covered *kushtakaas* or *gagiits* and their attested ability to swim powerfully, dive for shellfish, and so on, it might not seem too surprising to hear reports of sasquatches in southeast Alaska swimming to or from islands or past boats. One might suppose that the large number of cameras carried daily by cruise ship tourists might have made photographic evidence of swimming sasquatches more likely; alas, this hasn't happened yet. Alaskans, on the other hand, pass on the camera in favor of another tool—fishing equipment. In fact, anyone seen carrying a camera in Alaska is in effect identifying himself as someone who really isn't familiar with the place. On any summer day, a lot of people can be seen traveling through the waters of southeast Alaska. But come nightfall, almost all are safely ensconced in their seasonal apartment or onboard a cruise ship. Fishermen are among the few up before dawn or after dark and away from towns and cities. It is therefore not surprising that their accounts figure prominently in the stories that follow.

In the earliest of four southeast Alaskan reports of sasquatches seen in open water, John Green[1] made available a letter from a woman who had heard from her cousin and cousin's husband of a sighting near Ketchikan around 1960. The cousin's husband was named "Errol," and his story was passed on to Green from Nick Carter at Bellevue, Washington, through a letter. Green wrote that his letters to the man were not returned, implying perhaps the man had received them without replying. This is the story as Green received it in its original written form.

"She had a story to tell second hand, about a possible sighting her cousin [actually her cousin's husband] had in Alaska, 14 or 15 years ago. She saw the TV show of a month or so ago, went out and bought all three of your books and read them in proper sequence. When she came to the part about sightings in the Queen Charlottes and the bigfoot swimming, she suddenly remembered the story her cousin had told her...

"The story goes like this. Errol's father is or was a fisherman in Alaska and the boy spent his early years on boats with him. The summer Errol was fifteen, the fisherman were having a lot of trouble with something ripping up their nets and stealing fish and so on. Nobody knew what was doing this. Mrs. N. thought it was the summer of 1960 and the area near Ketchikan, not exactly sure. One night about eleven o'clock, the men were all in a shack on a dock, playing poker, when Errol's father remembered they had not pulled in their skiff and sent the boy to haul it up the side of their boat. Errol took a big flashlight and went along the long dock, apparently an L-shaped affair which ran parallel to the shore. He had to cross two other boats to get onto his. The three were moored side to side, with their boat closest to the beach. Just how far off it was, she had no idea, but not far, apparently, and the water was not too deep between the dock and the shore. This was salt water and very clear.

"The night was dark so Errol was using his big flashlight to work by as he set about hauling the skiff up the side of the big boat. He laid it on the deck, pointing inshore, and was bent down, hauling on ropes when he glanced up to see a human like figure standing in the water halfway between shore and the boat, just standing and staring at him. The boy froze for a few seconds and remembered the thing was not exactly a person but it had arms like a man and a head. It was probably wet and looked greyish color all over the head and body. It had round eyes, not big, beady like. It just stared at him, quite close, standing up to its waist in the water.

"When the boy unfroze, he screamed bloody murder and ran blindly, over the tied up boats, back up the ladder and toward the shack, still yelling his head off. The men came running, some with lights, and about thirty of them saw the thing. They shone several lights on it as it dived under water and swam away. They could see it under water swimming like a frog, arms forward over its head but not doing a crawl stroke. The legs kicked, the best description was like a frog. The men could see legs and arms as it swam out of sight. Nobody in the crowd had ever seen anything like it.

"Errol was so shaken by it that he never spent any more time on the boats or fishing. He wouldn't go near the place again. Mrs. N. said he would not talk about it much but he had once given her the whole story, with a sketch of the dock layout."

John Green[2] adds, "The last paragraph is not altogether accurate, since the lady has since told me that Errol, who is no longer married to her cousin, is now back in Alaska fishing for a living. There is of course no proof that the story is true, or even that there ever was a man who told it, but I don't have any doubt on the latter point." [S60A-map5]

A retired fisherman, about fifty-five years old at the time, also named Errol and living near Ketchikan, spoke with me in 1999. Errol told me that he knew about *ba'oosh* (Tsimshian for, among other things, ape and sasquatch) and admitted that some years ago he had seen something very unusual, but wasn't forthcoming about anything else. It was obvious in speaking with him that he was very upset by the publicity attending the previous story and sounded as though he had not wished the event to be made so public for some reason, rather implicating himself in the story in a backhanded sort of way. When I asked him directly about his possible involvement in the story, he delicately changed the subject.

I have interviewed a number of people in the village where he lives, having flown there once weekly to work for a year and a half in the early nineties. It was my feeling then, given the devout, serious-minded nature of many residents, that any young man reporting such an event would be subject to a certain amount of ridicule. This might not be an easy thing for a teenager, or anyone for that matter, to deal with, and there seems a strong possibility that this man was indeed the same young man who was said to have "run yelling his head off," in the report. If the dockside report is accurate, then the boy's response was one anybody might have experienced, and this negative ambience locally toward sasquatch reports in general may have kept a lid on a far greater number of reports than are documented.

This account certainly matches the description of sasquatches. While it would be difficult to prove its identity as that of a sasquatch with certainty, it does bear similarities to the aquatic behavior of sasquatches reported in five other cases in Alaskan water (besides the mysterious "sea-ape" documented by German

Figure12a: Kwakiutl sasquatch mask.

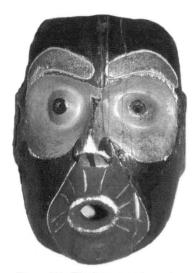

Figure 12b: Tlingit sasquatch mask, Haines Museum. *Photo: David Hancock*

naturalist Georg Steller on Bering's Alaskan expedition in 1847) and doesn't resemble any of the marine species known to commercial fishermen. (Steller's account has been debated for 150 years, but seems based on what appears now to be his first sighting of a fur seal, much different than the species he was familiar with.)

Reports of sasquatches swimming are not usual, but not rare by any means. One report of a sasquatch seen swimming was forwarded to me through e-mail by Oregon sasquatch researcher Ray Crowe[3], April 14, 1999. This was from well-known researcher Peter Byrne, who had interviewed a number of Seattle fishermen regarding "Sea Man," and moving deadheads that at close range turned out to resemble sasquatches.

As Mr. Byrne reported to Ray Crowe: "Tony Hawkins, on May 9th, 1990, was heading south from Petersburg [Alaska] as a guest on a commercial halibut fishing trip, and they were about a mile from land. He was on the bridge with other crewmen when one of them spotted a possible deadhead. As they approached it on the starboard side about 1:00 p.m. in calm seas, at fifty yards he noticed that the thing was upright above the water some four and a half feet, as determined through binoculars, and was about two or two and a half feet wide. It appeared to have hair like shredded bark, brown, and it had a head and shoulders. The arms were hanging straight down, he couldn't see any detail and he couldn't see eyes in the deep sockets. The face was rounded with a distinct nose and chin, some areas seeming swollen. The head was rounded like a rounded log end and the facial area appeared hairless with hair on top of the head. The face was a lighter brown, almost grayish in color.

"Thinking it might have been a carved log, like a totem pole, he noticed as they came abreast of it, it sank out of sight.....A log doesn't act that way....it would be bobbing, 'and it was looking at us the whole time, five minutes...it watched us the whole time.' Crewman

Eric said he had seen one quite a few years earlier, and again he saw one the previous September. Tony had read about one off the Oregon coast in 1948. Cameras were on board, but were stowed below.

"...The crewman on the ship, Lindsey Babich, notes that the ten-year-old vessel is a seiner and is 56.5 feet long. They were between Petersburg and Ketchikan in the Clarence Strait. Lindsey says that they had heard previously of the 'Sea Man', and it had been described exactly as they had seen it. He goes on to say that while heading toward Ketchikan [describing the same encounter], he saw the 'deadhead' noticing that it wasn't bobbing up and down vertically like a deadhead. With the binocs he said that... 'the thing had hemp-like hair, looked like a Sasquatch, had big eyes, a nondescript mouth, and resembled a piling with head and shoulders that went up to the head. Definitely there were eyes. They were large, almost as large as baseballs. They were black, like a seal's eyes but a lot larger than that...and he was staring right at us, just as curious as we were. The skin was a little like a muddy, tan-green.' They commented how a fisherman might dress up a deadhead and let it go as a gag, but to his mind, this was definitely a sea creature that he'd never seen before. It had a flat face, so no way was it a sea lion. And then it went slowly straight down, and didn't reappear. [S90A-map3]

"Another fisherman, Jay Morgan, saw 'sea men' twice, in two different areas, as Lindsey had also seen one twice. He says. 'I think they're out there in greater numbers, because fishermen tend not to look at these things. Jay Morgan knows two others who years ago were off Cape Flattery and they went right up to this thing and it scared them to death."

Given the number of reported sightings on ocean beaches, it is surprising that there are not more reports of sasquatches seen swimming. But of course, if they are primarily nocturnal, and can swim equally well underwater when boats approach, it would be difficult to catch them in the act. Then again, perhaps not. In a further excerpt from Peter Byrne, posted to an e-mail on *Bigfoot Digest* by Ray Crowe[4] on April 14, 1999, is a 1992 report of a sasquatch swimming, then diving, near the boundary of B.C./Alaskan waters in or near Work Channel, B.C., on the Inside Passage. With references to previous offshore sasquatch sightings also in the boundary waters between Canada and the U.S., the report goes as follows.

"Sighting of June 8th, 1992, 9:40 AM. Interview with Eric

Sowers. 'Rick and Randy saw it. I was down in the galley, I didn't see it. Supposedly, they were coming up real close to where they saw the one last spring, and they saw basically the same thing. They tried to get closer to it to take pictures and it went down and never resurfaced. It looked like a deadhead from a distance. When they got closer they could tell it wasn't a sea lion and wasn't a deadhead. They got to within about 150 yards, they said.

"They were doing about eight knots and the weather was blowing on the stern, steady in the water. The deadhead on the low range radar showed up. It never bobbed or weaved or moved. I had been doing some marathon wheel watches during daylight hours there for awhile, trying to see another one.

"The one that a guy and I saw last fall [1991], we were only 40 yards from. It was flat calm, not even a ripple on the water. About an hour and a half before dark, we're coming through Canada on our way south, right off Work Island. We were still up in the northern waters and I thought I saw a deadhead. I gave the binoculars to the guy with me and he thought it was a sea lion. The closer we got the more he said it wasn't a sea lion.

"I tried to take a picture, but it went straight down. It had been standing three to four feet out of the water and was steady as we got close to it, then it leaned over once to its left and it was facing us, it was turning to gaze at us. We could see two round dark spots that looked like eyes, but didn't see any nose or mouth, kind of flat. Color was dark tan, otherwise pretty dark brown. The facial area was a bit lighter. It had little bits of hair, not a lot, like the first one we saw did. The skin looked kinda wrinkled. We were 150 yards from land, it was between Work Island and us in a pretty narrow channel.

"Both of them, it kinda looked like there was shoulders. And the one these guys saw a couple of days ago, they said it was real pointed by the head, the top of it. But the one we saw last September was like shoulders about half way up it, maybe a foot below the top of it." [S92D-map6]

This is the most recent of four reports of creatures resembling sasquatches seen swimming or, more properly, treading water while observing a boat, then diving. There is some confusion in dates regarding the U.S./Canada boundary waters sighting as documented above, but this is apparently due to other similar nearby sightings the various crew members had mentioned to each other.

The preceding reports suggest that not only do sasquatches swim, they do so powerfully and without the typical primate pattern of "dog-paddling" or the use of arms as in human swimming. Apes lack the range of shoulder joint mobility to use the arms in a human swimming or overhand throwing motion. This style of swimming described for sasquatches would accord well with a more apelike anatomy and may be further evidence of a nonhominid ancestry. A more apelike ancestry might well include behavior common to all the great apes, such as curiosity toward man and his artifacts, including boats and vehicles, as we shall consider next.

1. Green, 1978, *Sasquatch: The Apes Among Us*, Hancock House Publishers, Surrey, B.C., pp. 430–432.
2. Green, 1978, *Sasquatch: The Apes Among Us*, Hancock House Publishers, Surrey, B.C., p. 121.
3. Crowe, Ray, noted in *Bigfoot Digest*, April 14, 1999, <raycrowe@aol.com>.

5 Trouble with Sasquatches

Trouble on Prince of Wales Island

During the 1950s, around the small village of Klawock, on the west-central coast of Prince of Wales Island, a number of strange stories began to circulate within the community. Mr. A. K., a dark-haired, retired Ketchikan man in his sixties, presented a personal account of what sounded like perennial "sasquatch harassment" of his and related families in Klawock during that time. He told his story to me in 1998.

"The consensus among the M. family and our family was pretty much the same about all the *kushtakaa* happenings. I was a young man in Klawock at the time. They had trouble with one or more of them right up to the late fifties. I heard most of them but there was one I was involved in one fall, sometime in between 1952 and 1955, I believe. I was staying with relatives at the old M. house on the edge of Klawock Lake. A *kushtakaa* was fooling around outside one night trying to scare us at the M. family home. Our cousins were there, and there were about six family hunters rounded up to put a stop to it. It would go round the house in the bush, faster than you could make your way through it out there. It would just crash through the bushes real fast making crying sounds or just hooting and breaking branches making a racket. I remember it coming around and shaking the door knobs; but it never turned 'em, just rattled 'em and shook some metal gas cans that were beside the roof.

"We got the guys who were good hunters rounded up, but when you'd shine the light, there'd be nothing there. Everyone went back to their own homes, except my uncle W. K., an army vet who could really throw a big knife, and a couple of family friends. Then it came back and started rattling the door again, and Uncle W., he had been in combat, he took off after it and you could hear them running, hear his boots leave the cement by the house and chasing it down towards the beach. He came back, saying he had thrown his knife and 'heard it hit something, but the *kushtakaa* didn't stop.' In the morning we went down to where he heard it hit and found his knife buried up to the hilt in the telephone pole that stood between the house and the water. Another thing we tried to stop him with was to run out a whole lot of halibut gear around the house. He must have snagged up at least once because some was tore up and there was coarse black hair all over." [S52A-map4]

There is a curiously similar report from a relative of the same family, a senior Native Ketchikan woman, who in a 1999 interview, asked simply to be called Miss M. She stated that as a girl she had been living in Klawock at the time, and her family had been visiting relatives at the house in question. She said that she had heard of several stories from 1952 and onward in which she was not involved. But she added: "There had also been a previous encounter while I was staying there in the summer of 1951 [at the same house] with the gas cans up by the roof and my relatives had heard the creature on the roof scratching the metal. One man and a few fellows chased it down into the lake earlier that year, a big black one, but it just dove into the shallow water and swam away, they could see it kicking its legs like a frog. 'It could swim real fast, it never used its arms,' the men said.

"It was about seven feet tall, and there must have been two of them, because that one left reddish brown hair all over the roof. The next day there were long scratches on the metal roof and we saw reddish brown hair all over the place up there. It was quite frightening. Everybody knew about it though."

The woman added: "All my older relatives saw it dive into the lake another time as well—that was about the time they tried to catch it at night by setting bear traps all around the house. In the morning we all saw the traps had been sprung during the night with sticks in them. I don't know if that was the black or the reddish one." [S51A- map4]

These incidents are known to almost every senior member of the small town of Klawock. They are seldom discussed, however, due to fear of ridicule, as well as the belief that any given *kushtakaa* might be the embodiment of a departed relative. There is probably more to these related stories, but just as in the Haida community of Hydaburg, an hour or two south down the road from Klawock, encouraging older Native people to share stories with an outsider can be difficult, especially considering the tribulation some people have gone through in regard to their traditional beliefs in these creatures.

The following two Prince of Wales Island sasquatch reports are interesting because they both describe what sounds like a form of moderately aggressive behavior toward moving vehicles. Even a secure vehicle may not provide immunity from fear in some alleged sightings. Following is a 1998 interview with a Ketchikan woman who shared her frightening experience.

"One night in July, 1984, I was driving with two other people from Hollis, Prince of Wales Island, to Thorne Bay, also on the east side of the island. It was late, around eleven o'clock, and we were heading up a long straight stretch of highway less than ten miles from Thorne Bay. As we headed along up a hill with my driver's window down, I caught a whiff of some awful smell and rolled the window up. Right then, we noticed a tall figure running just along the shoulder of the road beside the car—running like a man. We were doing about thirty-five miles per hour. In the edge of the headlights just fifteen feet to my side, we could see a creature about seven feet tall and covered all over with dark hair. It was swinging its arms like a man but was a lot heavier built. The slope of the road was about five to ten degrees I would guess, and it kept up with the car for about ten seconds. It didn't make any sound or anything that we could hear, but I have to tell you I was scared. We sped up and it seemed to veer off as we left it behind. We didn't stop till we got into Thorne Bay later that night. We didn't tell anyone because you can imagine what they would say. That was the only time I ever saw anything like that anywhere." [S84A-map4]

Sasquatches have been documented elsewhere as running at speeds of up to forty-five miles per hour, perhaps faster, but no one else in Alaska has reported timing one in a vehicle. Two Hydaburg women recently reported something knocking on their driver's side window as they drove from Hydaburg to Craig, Prince of Wales Island, one night years ago. Unfortunately, neither of them had wanted to take a good look and neither can recall how fast they were traveling on the (then) gravel road. "It was not an injured bird," they said. If they were correct that it was not a bird, about the only thing capable of performing such a feat would be a sasquatch. Olympic athletes reach speeds in the neighborhood of twenty-five miles per

Figure 13: Sasquatch reported running beside a car west of Thorne Bay, Prince of Wales Island, AK, in 1986. Sketch by author under direction of the witness.

hour and therefore the extra foot of stature usually accorded a sasquatch might easily account for a running speed of thirty-five miles per hour, even for a mature, heavier sasquatch.

I have heard of two other incidents allegedly involving sasquatches and vehicles in southeast Alaska, one in the nineties in which the back door of a camper was torn off while the occupants were driving at night to Hydaburg. The other incident involved three Ketchikan teenagers claiming a hair-covered creature had jumped onto the roof of their Volkswagen near Harriet Hunt Lake, Revilla' Island, in the late seventies, squashing it down, then jumped off and ran on two legs into the bush.

While these are far-fetched stories to say the least, I have been able to confirm at a body shop that the Volkswagen did indeed have considerable roof damage with no evidence of having rolled. A Ketchikan sporting goods retailer also told me he and a friend had pulled up beside the unhappy trio only moments after they had brought their vehicle to a stop below a rock cut. He made the observation that he thought it unlikely that the trio had all jumped on their own roof to squash it down, and that their story seemed as likely to him as any other explanation.

At least three nocturnal incidents describing cars picked up by the rear bumper and shaken at Ward Lake on Revilla' Island are noted in a later chapter, and in at least one case, an occupant thought she had seen a figure resembling a sasquatch shortly before the event. This latter phenomenon has been reported in association with sasquatches for vehicles, trailers and even houses in British Columbia and northwest states, but has not been generally reported anywhere else in Alaska.

Trouble on Revilla' Island

Perhaps, like small boys, sasquatches just seem to have a fascination with cars and trucks. Five reports from southeast Alaska list vehicles as the object of a sasquatch's attention. In three of them, the location was given as the old, formerly unpaved, parking lot at Ward Lake, the focal spot of numerous other reports, most of them sightings.

In a second-hand report given in 1999, Mrs. Brenda R. of Ketchikan, stated that around 1988, her neighbors, a certain Watson family, had a strange incident with their car at Ward Lake. She believes the event described took place "in the summer of '88, as the neighbor family came home from a trip to Ward Lake."

"I watched as the neighbors all got out of their vehicle in their driveway, and the mother was discussing what they had seen crossing the Ward Lake parking lot toward them as they had been sitting in their car the previous day. Apparently both parents had been in the front of the car and had seen 'a large creature walking on two legs across the parking lot in the near darkness and proceed to walk along the passenger side of the vehicle, scratching it' as it did so."

What impressed Brenda was, "the clearly visible row of five long, parallel scratches from one end of the passenger's side to the other." Mr. Watson apparently worked at the time in the Coast Guard, but has since retired to Florida, and I have been unable to locate him for an interview. There is not much one can say about a report of this nature and it certainly leaves one wondering about what else might have caused the five parallel scratches as described. [B87B-map5]

While bears certainly leave claw marks and long scratches on things, it is difficult to envision a bear walking along the side of a vehicle to do so. It is rather the type of behavior noted at the same exact place at Ward Lake in two other instances, where a parked vehicle was the object of violent attention (1966 by Ms. D. Y.; 1974 by Ms. L. W.). Another sighting occurred less than 100 yards away at a picnic shelter. One thing surprising about these three reports, all at the same exact location, is that they span twenty-three years. The three reports are quite spread out in time for any one individual animal, but it is surmised that forty to forty-five years is not excessively long for a great ape to live. Then again, it is possible that there was more than one creature involved. Whether this behavior might represent sasquatch aggression or simply idle curiosity is difficult to say.

6 Sasquatches on the Shore

North of Revilla' and the Cleveland Peninsula

Although sasquatches are commonly reported in or near the ocean, nearly all seem to describe male creatures and females are almost never seen. An amateur Ketchikan taxidermist, Mr. M. M., reported in a series of interviews in Ketchikan, 1999, an extraordinary sighting in October, 1992, of a female sasquatch halfway between Wrangell and Ketchikan from a skiff. He admits he contemplated shooting the creature. Mr. M. is a very proficient hunter and taxidermist of deer, moose and big game. While he did not wish to possibly jeopardize his successful business endeavors with disclosure of his full name, he was interested in sharing what he had observed. He did not seek any publicity or remuneration and was most patient in subjecting himself to long, repeated interviews in which he consistently confirmed the details and succeeded in avoiding deliberate "traps" when discussing anthropological details such as foot anatomy. His full story unfolded as follows.

"In 1992 we had been up Ernest Sound [on the top of the west side of the Cleveland Peninsula] crabbing and it was October. The main boat was anchored in Sunny Bay, and I had taken the little outboard skiff out for an early morning trip south around the points to take a look around and possibly get some meat for the galley. I had a .243 [rifle] with a variable nine-power scope and was following the shore real close. I had been moving south for an hour with the little outboard scouting close to the shore, going slow and as quiet as I could. I had just come around a sharp point into a bay and was about a hundred yards offshore, when I noticed something black on the shore and killed the engine. The echoes of the engine died away. I

71

couldn't hear the sound of the ocean from where I had just come and I kind of figured I had been lucky and this "bear" hadn't heard me. I lay down flat in the skiff and looked through the scope at the back of the animal, which I still thought was a bear hunched down on the beach. The only things sticking up over the gunwale were my head and the rifle, and I turned the rifle-scope power up to nine power, filling the scope with the creature's backside. It was hunched over in the sand by a little creek, about 140–150 yards away.

"The moment I saw it was sitting on its heels, I knew it wasn't a bear. It was a very big manlike creature, and around two minutes or so—maybe I made a noise or something—it turned and looked at the boat for a moment. As it turned its shoulders and upper body to look sideways at the boat, I could see it was a female. The face was flat and dark-skinned; she had a broad nose with flared nostrils. The face area had little or no hair on it. Laying there in the boat, I watched the whole creature in detail, just sitting there.

"At first I held my fire because the creature very clearly was not a bear. She was squatting on her hind legs doing something in front of her with her hands. The soles of her feet were apelike or manlike. As she turned her upper body and head to her left, the right heel came up and the whole length of the foot seemed to bend inwards, very strange, it almost seemed to me to make the letter 'C.' It seemed much more flexible than a human's. The head was somewhat pointed and the face had dark skin, it was like an ape, sort of. The body was thickset with a very thick neck.

"I could see it was a female by its breasts, which were almost completely covered by black fur. The fur on the body was about four inches long. She sometimes had her hand to her mouth. If she was eating, I couldn't make it out. But the face skin was dark brown or black and wrinkled, with strange little patches of pinkish lightly pigmented skin. I couldn't see teeth. Her eyes were startling—I could see light brown eyes like a doe's with large, dark pupils when she stared at the boat for a few seconds. She didn't seem concerned with me at all. I don't think she could see me. She had a wide mouth but no really obvious lips. The chin was narrower. There was no hair around her mouth, nose or eyes. I think I mentioned her nose was broad and her nostrils were flaring, like some people's. The hair on the back of her head and shoulders was longer, maybe twelve inches, and it seemed to run down to between her shoulder blades in sort of a mane.

"She turned back to the thing she had been doing. She wasn't human, that's for sure. Her hands had nails that were dark brown. After six, seven minutes or so she got up. She didn't use her hands. She just rose straight up on her legs and I could see she had a huge abdomen, bulging from her hips up high to just under her breasts. She looked pregnant. Her breasts weren't floppy. They were about the size of dinner plates. She was about seven to eight feet tall but really broad across the hips, buttocks. My guess was over fifty inches. She looked healthy, in good shape.

"I thought of squeezing the trigger a good number of times," he added. He then confided to me in all apparent, if somewhat guilty, sincerity, "I wondered if I couldn't get some scientist to pay a good price for her. Do you have any idea how much that might be?" [I demurred and returned to interviewing.]

"But she was looking pregnant and besides it was just a beautiful animal. I could have squeezed [the trigger] ten or twelve times. I had the cross hairs on her head, at the temple. That would sure settle the scientists. But anyway, I just didn't. 'I've seen one now,' I thought, 'and that's good enough for me.'

"When she started to walk to the trees, she walked with a slight forward slump and a kind of waddle from side to side."

Following is a question and answer session between myself (Q) and Mr. M. M.

Q: "What did her feet look like then, did you see?"
M. M.: "Well, nothing too much, but she had big heels, they kind of stuck out."
Q: "How long did you watch her?"
M. M.: "About ten minutes in all, until she went into the trees."
Q: "Until she disappeared from view?"
M. M.: "Yes."
Q: "Did you hear anything?"
M. M.: "I thought that when she first turned to look at me in the boat, she made a grunt."
Q: "No screams or anything?"
M. M.: "No."
Q: "And you left?"
M. M.: "After she walked into the trees, I continued on in the skiff."
Q: "You didn't go onshore?"

 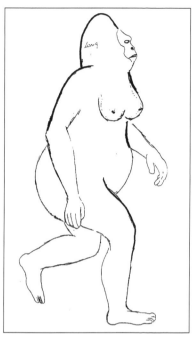

Figures 14 and 15: Eyewitness illustrations of face and torso of a female sasquatch, reported by Ketchikan taxidermist M. M. on a beach in Ernest Sound, AK, 1992.

M. M.: "Yes, as a matter of fact I stopped on the way back to look for tracks."
Q: "Did you find any?"
M. M.: "Yes, I found a couple in the softer sand where she had squatted. They were about fourteen to fifteen inches long, by seven to eight inches wide. They were about one inch deep."
Q: "How deep were your tracks?"
M. M.: "Not a quarter of an inch."
Q: "Did you smell anything?"
M. M.: "No."
Q: "Did you notice any other things?"
M. M.: "Yes, well, her head was shaped with a slope to it. It went up about four inches vertically above the ridge above her eyes and went back about ten inches, sort of rounded off back there."
Q: "What was the most forward part of her face?"
M. M.: "Her mouth."
Q: "Not her nose?"

M. M.: "No."

Q: "You've done taxidermy on bears, how heavy would you say she was?"

M. M.: "Well, I don't know exactly, but, compared with a bear, well, at least 500 to 600 pounds, maybe more."

Q: "Any other final thoughts?"

M. M.: "Well, you know, sometimes I thought I should have shot it."

Q: "I'm sure that afterwards, from time to time, any man who has hunted would feel that way, but you won't ever need to wonder what you might have ended up feeling like afterwards if you had."

M. M.: "No, I suppose not." [S92C-map3]

This is the taxidermist's story, make of it what you will. In three follow-up interviews, he did not change any of the details. He was able to provide me with several additional comments on such features as the overall body hair, which he thought was about four inches long. (The mention of a "mane" is also similar to that noted in Dr. Fahrenbach's previous report from Ed L. in Prince William Sound in the early 1990s.) His comment on the peculiar projecting nature of the heels, something only well known to anthropologists who have studied the frames from the Patterson film from Bluff Creek in northern California, is rather remarkable. It may be worth noting that independent analysis of hundreds of tracks by physical anthropologist Dr. Grover Krantz has led to the same projected heel anatomy, based on a nonhuman model of ankle anatomy that would require the sasquatch's ankle to be placed further forward on the top of the foot than in humans. The broad (seven- to eight-inch) width of the foot, approximately half of its length, is close to the maximum 2:1 length to width ratio of casts of sasquatch tracks as documented by Krantz[1]. Krantz's book *Big Footprints* had not to my knowledge been made available anywhere in Ketchikan at the time of this publication, and it is doubtful that M. M., a hard-working family man with several jobs, would have had access to that literature or the time to read it. In a follow up interview with John Green and myself in the summer of 2000, Mr. M. was able to recapture all of the points made in the first interviews.

Green came to Alaska in the summer of 2000 and wanted to make his own opinion of the account, intrigued by the man's story.

Following the interview, Mr. Green confirmed my impression of credibility, commenting, "It was certainly worth hearing, he seemed sincere, and the observations of the flexibility of the feet match with what we have come to hear in more recent reports." For his part, the man seemed not to have even been aware of Green's authorship, research or books at the libraries in town. In fact, he had even been a bit hesitant in coming forward in the first place, since local southeastern Alaskans are a bit shy of discussing anything reminiscent of the fabled Tlingit *kushtakaa*.

The taxidermist himself was surprised to see the volume of sasquatch reports from southeast Alaska shown him at the follow-up interview, saying he had not been sure about coming forward. He seemed quite relieved to have gotten it off his chest and see that he was not alone. Perhaps the most amazing aspects of his account are the two eyewitness sketches he drew of the female, one of which is a side profile of the entire body as she walked away, the other a detailed illustration of her face, both drawn in 1999.

James Nanez, a Ketchikan ferry worker and avid hunter, described an unusual creature he noticed watching him while bear hunting east of Ketchikan on Revilla Island in June, 2000. James stated in an interview in August, 2001: "I was spending the morning of June 8th, last year, scoping out the forest around a muskeg at the head of Salt Lagoon at the north end of George Inlet. I was sitting by a big rock and log and was looking through my scope about 100 yards away at the bush around large rock, when I noticed a big black figure squatting on the rock, just looking at me. I thought, 'What the heck was that?' I hadn't noticed it before because its outline was broken up by the foliage. It was almost black, had a dark face and was just watching me. I had first noticed it with my scope at low power, then with my eyes, and then looked again at high power. It was still staring right at me. It was big, about seven to eight feet and really heavy, maybe six or seven hundred pounds. As I was holding my rifle up again it stepped down off the rock and started to back away from me toward the creek. But it never turned its back on me or turned its face. I stood up to keep it in view, and it just kept backing away, never taking its eyes off me. I think it was a male, I could see all of it at times but it was keeping a low posture as it backed away. While it backed up, it was bent in a crouch like a linebacker. Even crouched like that, it was about five feet high. I do remember that its hair was about four

Figure 16:
Author's illustration of a sasquatch reported by bear hunter James Nanez at head of George Inlet, Revilla' Island, AK, 2000.

inches long on its shoulders and arms. It didn't make a sound as it moved, and I watched it slip out of sight into the cedars and hemlock on the near side of a creek. I was moving towards it with my rifle scope trained on it when I saw it last and eventually got up to where it had been. I found only one clear track with five toes, about sixteen inches long by maybe five and a half inches wide. I don't know if it was more human or ape, I still wonder about that, but it sure was a fascinating thing to watch." James indicated clearly that at no time did he take it for a bear or even contemplate shooting it. [S00B-map5]

A creature the size of a sasquatch would likely have to find calories wherever it could. Based on reports, it appears they have an appetite for almost everything. As in the 1992 Larry Thomas report near Ward Lake, the following account describes a sasquatch tearing at the bark of a tree. A burly Ketchikan hunter, Darrell P. said the following in an interview in 1995.

"On a Sunday in October of '94, I was deer hunting with a friend one mile northeast of Saxman, right here on the [Revillagigedo] Island. Kevin and I had been up in the muskegs about four or five hours, it was good weather, not raining and it was about 6:00 p.m. We were beside a muskeg, in some pine and spruce, maybe eight to ten feet tall but with lots of big old logs and salal bushes. I was standing on the end of a big three-foot diameter log, looking around, and Kevin was lower down to my right about fifteen feet away. Through the branches, we saw patches of what at first appeared to be a bear with grayish brown fur standing up thirty or so feet away.

"I could see a dark brown head about seven feet off the ground tearing the bark off a small pine tree. But as I watched, I could see it was clearly not a bear. The left hand doing the tearing had five digits, and the skin on the palm and underside of the fingers was a real pale gray color, not dark or black like a bear's. The skin on the parts of the fingers where the hair started looked darker—a brownish gray. I could only see part of the head and the one left hand. The fur or hair on the back of the hand was a dark brown, about two or three inches long. The jaw looked like a large human jaw, and he appeared not to notice us. I motioned for Kevin to get back on up where I was, but he was just as busy motioning me on down. I think I must have made a noise because suddenly it moved out of sight toward the other end of the big log. It may have heard me because it made a sound like a grunt.

"It was real quiet, the wind was from him to us, we smelled its wet fur. The next thing I knew I felt something hit the log, the log came straight up and I was catapulted off the log to one side, landing in some salal. I got up unhurt, Kevin said, 'Let's go!' and I agreed with him this time.

"We headed out of the muskegs and back down to Saxman. Kevin and I estimated I flew about twenty feet from where I had been standing on the log, and at the time I weighed 250 pounds. I can tell you that whatever stepped on the other end of the log was big.

"I remember catching a glimpse of him as he moved from the pine tree, he took a stride and turned. I would guess his weight at about 800 pounds."

He concluded by stating that they had seen no tracks other than numerous deer tracks, no bear sign or any other sign, sound or odor. [S94A-map5]

Darrell added that he had not the opportunity to get back hunting that particular part of Revilla' Island since, nor had he told anyone else this story for fear of ridicule. His partner corroborated the basic facts as related, but indicated to me he did not want to discuss it further. What might be represented by the alleged tearing of bark is speculative, but one could infer that the sasquatch was foraging for grubs beneath the bark or possibly for the nutritious cambium layer beneath.

Prince of Wales Island

In an unusual report from southern Prince of Wales Island, Mr. R. S. of Ketchikan, a retired chemist, reported a creature he and his daughter had seen from a sailboat around 1975. In an account given to Alaska's Southeast Sasquatch Research Group of Ketchikan in 1999, he told the following story.

"We were sailing halfway down the south arm of Chomondoley Sound on the east side of Prince of Wales Island. There was no sound. We were sailing north toward a peninsula on our right, on the east side of the sound. Suddenly my daughter pointed to the cobble beach and exclaimed, 'There's a black bear on the other side!' I handed her the binoculars while we cruised past. Our speed would have been about two knots. She handed me back the binoculars, and I saw a chestnut colored creature, reddish brown, down on all fours. In that posture it appeared to be about three to four feet tall. It had its rump toward me and I took a good look at the hind legs. They were long. I looked at the head and could see no ears visible. It was moving along the cobble beach up toward the trees. It was bent over and moving slowly, not bothering to turn and look up, of course the sailboat was quiet. I had the impression of legs longer than a bear's. We went by again later but it was not there. It didn't look like a bear to me!"

Although at first glance this might seem to match the description of a black bear, the reddish color variation is quite rare for *Ursus americanus*, even in the bear's brown color phase, and the noticeable lack of ears and the unusually long back legs all suggest this may have been a sasquatch foraging in a quadrupedal posture up the beach. [S75A-map4]

Also on southern Prince of Wales Island, logger Mike Stough of Wrangell reported a beach sighting of two sasquatches he had in winter early in 1980. He was flying into Hessiah Inlet, fifteen miles south-southeast of Hydaburg on the west coast of the island, to meet his father, Mr. Dick Stough, and two other men for beach logging work. Mike gave this interview in January, 2000.

"I was flying in on Weber Air, and the pilot had come down to circle the point so as to land inside the inlet where our boat was. It was about two-thirty in the afternoon, late January or early February, overcast with a very light drizzle. We had seen no other boats and, as we flew low over the beach, I was kind of curious to see two black figures down there, bent over at the waist, stacking up rocks. They were right beside a big cedar log and were as thickset through the chest as the log was wide. At the time, I thought they were maybe two of our guys because no one else was around, but I didn't think our crew had any black rain pants and black jackets with hoods.

"They didn't wave or anything as we came past and over the point to touch down beside the boat, tied up right near shore on the inside of Hessiah Inlet.

"The first thing I said to Dad and Bruce as I stepped down was 'Who the hell you got on the beach?'

"They just looked at me for a second and then replied real surprised, 'Where on the beach?'

"So I told them, 'just a couple of hundred yards around the point.'

"They replied, still surprised, 'No one!'

"So I said, 'Well, who are those two guys in black rain suits?'

"When we got around the point to check, they were gone, but we could see where they had been digging around by this big two-foot thick cedar log, and there were rocks piled up like someone had been looking for crabs or something. There on the beach gravel were two sets of two-legged tracks, one fifteen and a half inches long and the other thirteen inches long. The pilot hadn't seen the figures as he was busy flying the plane, but I guessed, judging by the log they were standing beside right there, that they would have to have been about six and one half to seven feet tall and have weighed at least 500 pounds. The tracks weren't real clear. They were in overturned cobble just below the high tide mark. The forest above the beach was hemlock and cedar. We didn't see anyone outside our party the whole trip." [S80A-map3]

The details of this report have been confirmed by the other men mentioned and they all concur. What Mr. Stough described does not sound at all like two large bears. And "men in black" with no boat in such a remote spot seems unlikely, to say the least. Sasquatches, on the other hand, seem to be rather fond of Alaskan beaches, and whether the creatures reported were looking for crabs, digging out clams or just marking territory, any of the above behaviors would match other sasquatch reports. Interestingly, this is exactly the subsistence pattern ascribed in Haida folklore to their sasquatch, the *gagiit*, or "land-otter man" (see the chapter *Native Sasquatch Folklore: Haida*).

Annette Island

Point Davis, a scenic spot on the southwest tip of Annette Island, was the location of a surprising sasquatch sighting late one July around 1970. Haylee Nix, Hydaburg, recalled in a 2001 interview, that late one day, she had been camping with her dad, uncle, her two sisters and friends, enjoying an evening at the picnic tables beside the ocean.

"I was fourteen years old at the time. We were all around the picnic table less than 100 feet from the ocean and had a roasting fire. My sisters were there, playing around the table. We had thermoses, blankets, toys and all the picnic stuff. It was about nine p.m. and there was a clear sky. I was sitting on top of the wooden table and had just looked up toward the ocean when I saw a tall dark figure walking along the water's edge, just yards away. 'Look at that thing over there!' I called out. In an instant the whole family was gathering stuff up and running for the cars. Everybody was grabbing up stuff and running around. I remember just staring at it from where I was on top of the table. It was about eight feet tall or larger and was just walking to the right along the waters edge. It was dark brown or black in color, heavily, built, and walked like a man, swinging its arms. It never even looked at us. I could see it was covered all over with long hair. It had big feet, they were covered with hair. It was walking on two legs all the time, walking fast. I remember its hair was long and seemed to hang. I jumped off the table and hurried with the others to get in the car and we left in a hurry. I never really told anybody about it." [S70A- map3]

The woman's description of the creature matches that of most other reports, especially in its color and an apparent indifference toward people, with only the reference to greater size and long hair setting it somewhat apart from most sighting reports. The behavior of the family (confirmed by several relatives who alleged to have been present at the time), retreating from the sasquatch, is both typical and quite understandable.

While most road sightings last less than ten seconds, beach sightings seem to last a bit longer, and the following report is typical. Mr. Dan A., a hardworking, soft-spoken Tsimshean fisherman in his forties, told me in an interview in 2000 of a brief encounter that he and his father had with a *ba'oosh* on Annette Island. In the fall of 1992 the two men had been deer hunting at Crab Bay on the secluded east side of Annette Island. It was evening, and they were looking up from their camp chores at a point on the other side of the bay about 200 yards away.

"I suddenly noticed a tall 'stump' that had started to move along the edge of the bush. I called my dad's attention to it and he saw it too. At first he thought it was a wolf, and said so, the color was not right for a bear. It was a light brown; I guess you would call it a beige color. But it was upright, about the size of a man, and after thirty seconds or so paralleling the beach, it disappeared into the bush. We saw no deer there but the next day we did see a wolf nearby. But wolves aren't beige, man-sized or walk upright." [S92B-map5]

It is rare to hear of infant or juvenile sasquatches by themselves. But Ms Tina D. of Ketchikan stated in an interview in 1996 that she thought she had seen something of this sort on Annette Island one summer around the early 1980s while beachcombing with friends. Tina is a Tsimshean, dark-eyed and shy, but with an artist's talent for detail. Here is her 1996 interview.

"I was walking the beach south of Metlakatla one summer morning in the late seventies or early eighties, just following a minute or so behind some friends. I had stopped to look for eagle feathers and whatnot on the beach, when a movement about forty yards away at the edge of the forest caught my eye. Just coming out of some thick salmonberry bushes which bordered the beach, was a small, black, fur-covered figure about three to four feet tall. My first thought was 'What's someone's child doing in a black snowsuit in the middle of summer?' It was warm and sunny out, and it didn't make any sense at all. It was covered in black fur from head to foot. It had just stepped out onto a narrow dirt road, which cuts through forest and salmonberry bushes between the beach and a road inland which heads south of town. The little figure seemed to walk with some difficulty,

Figure 17: Sketch of a small sasquatch reported on a beach at Annette Island, Alaska, early 1990s. Illustration by author under direction of witness.

a lot like a human child does when it's a toddler, but this was bigger, about three to four feet tall. It suddenly seemed to trip on something, and pitched forward onto its face, throwing one arm up over its head as it fell. It didn't cry out or anything, there was just the 'whump' as it hit the ground. Right away, it got up a little unsteadily and just continued walking into the salmonberry bushes on the other side of the old dirt track. I didn't hear or see anything else after that, and just hurried on up the beach to join my friends." [S90B-map5]

This report is not very clear that it was a sasquatch that was seen, but the details seem to fit that description more easily than a human child wearing a black fur suit in the summertime. If so, what was an infant sasquatch doing alone? Or was the mother simply nearby, watching? That it did not cry out as it fell is not indicative of much; many rural children, native and otherwise, are just as stoic when it comes to scrapes and tumbles, but the "toddling" gait does not fit a human child of over three feet in stature, unless it was an awfully big kid. Toddling is seen in two-year-old humans, but for a two-year-old hominid to be three feet in stature, he would have to attain an adult height of eight or nine feet. This again suggests that what Tina D. reported may have actually been a young sasquatch. If she saw anything at all, there is not much else it could have been. Until a better explanation is offered, this account accords quite well as a sasquatch report.

Another sasquatch, much larger this time, was also reported in broad daylight on Annette Island, around the same year. Commercial fisherman Joe B. is a personable, hard-working Ketchikan man, with a soft-spoken manner. In 2001, he reported that on a sunny summer's day in 1981 he had been transiting Nichols Passage southbound, a channel that separates Annette Island to the east and Gravina Island to the west.

"I was heading south around the north end of Annette Island, and was passing a point on the shore called Cowboy Camp, a deserted village site. Just as I was viewing the shore several hundred yards away, something tall stopping on the shore caught my eye. It was dark brown, maybe black, and was about eight feet tall judging by the rocks, logs, trees. It was really heavy built and had just stopped walking to look at my boat as I passed by. When it turned after a moment and walked into the trees, it walked just like a man, but it was covered from head to toe with dark brown or black hair. I couldn't make out its face or anything, but it was the same color all over." [S81B-map5]

South Alaska Mainland

In several interviews in 1994, Mr. Tom B., a Ketchikan ferry worker, gave an intriguing report of a sasquatch digging on a beach close to the U.S./Canada border. In a matter-of-fact manner, Tom recalled his amazing sighting.

"It was in '67 or '68, late summer, and I was out of school. My dad and I were with this power troller, up there on the Portland Canal. We were just messing round setting up shrimp pots. We were about halfway up between Tombstone Bay and the next bay up on the American side.

"That morning we were anchored off this small anchorage close to shore, and my dad's friend woke me up to see something on the beach. I looked through the binoculars. What it looked like to me was a manlike, dark, hairy creature, digging in the sand like he was digging for clams using a stick. It was just after dawn, the days were longer you know, so there was still plenty of sunlight, but it was cold in the morning.

"We were about 200 yards from shore, which was a shallow, protected area. There was a creek coming down with a little hump of silt on both sides, goose tongue plants growing beside that and thick forest up from there. The creature was on the right-hand [north] side of the creek. The creek would have been about the same size as White River north of Ketchikan—about twenty-five feet across.

"Looking through the binoculars, I got excited and went down to wake up my dad. By the time he came up, the creature was walking on his hind legs back towards the left and into the forest. I would say the total time I observed him was three to four minutes. As my dad was looking through the binoculars, I could see with the naked eye that it was upright, walking on its hind legs, striding like.

"I don't know how big it would have been, six or seven feet maybe. While I had watched it digging, it had been squatting, digging with its right hand. The stick it had been digging with would have been about the size of a small baseball bat or halibut club. While digging, it would pick something up with its left hand and, it seemed like, would put something to its mouth. It didn't make any sound, but I think it heard us moving about the boat and then walked off. That was the last we saw of it." [S67A-map6]

Mr. B.'s account is somewhat unusual in that the creature reported was either not aware of the people watching or not unduly disturbed by the anchored boat. When it did eventually react however, it did so in the manner typically mentioned in most reports, avoiding further contact. It seems it was sufficiently preoccupied with the task at hand, could not hear over the sounds of the seashore or perhaps simply used to seeing boats anchored overnight. This is not the only report of its kind. A commercial crab fisherman described almost exactly the same behavior off the Cleveland Peninsula in 1991.

More than one southeast Alaskan hunter has reported having a sasquatch in his rifle sights. The late Terry Wills of Ketchikan forwarded one such sasquatch sighting by the late Mr. "Tink" Durgan of Ketchikan, near the extreme southern tip of southeast Alaska in about 1979. Mr. Durgan had related to Wills and others that he had been on his boat at the head of Hidden Inlet, off Pearse Canal, just fifteen

Figure 18: Illustration of a sasquatch reported digging on beach on Portland Canal, circa 1967, by witness Tom B. Sketch by author under direction of witness.

miles northeast of the Canadian border. Following is the story as related to me in 1999 by Mr. Wills.

"Tink told me shortly after that time that, one day around 1979, he had been on his boat near the head of Hidden Inlet. He was at a spot where by looking up he could still just see the alpine tree line. He was cruising slowly near shore, just a couple of knots, when he saw a heavy set, black figure walk out of the trees above the beach. It walked toward him and just stood there, so he stopped the boat and looked at it over his rifle barrel. 'It was not that far away,' Tink had said, 'fifty to seventy yards at most.' As he looked at it, he could see that it was real heavy, six and a half, maybe seven feet tall. Tink had said, 'It stood like a human,' and had added, 'I couldn't shoot.'"

Both Mr. Wills and others who attested to hearing Mr. Durgan's story affirmed that Tink seemed sincere about his encounter. [S79A-map6]

There are still good runs of salmon at Hidden Inlet, the site of an abandoned cannery, and plentiful blacktail deer as well. It is about as remote as one can get on the coast, which is remote to begin with. A few boaters may venture there in July or August, but, as with so many other inlets along the coast, the closure of the old salmon canneries has taken the last remaining families with them. The forest quickly closes in with hemlock, red cedar and moss. In the daytime, the eagles survey the rocky shores; in the evening, bears and deer descend the creeks to the beach. No one watches them. There are many such abandoned canneries or camps on the map; if a sasquatch were to pass by one today, even leaving its tracks in broad daylight, there would be no one to take notice.

Just a few miles north across the peninsula from Hidden Inlet is Mink Bay. As in the previous beach sighting, "watchful" seems to describe the behavior of the sasquatch in the following second-hand report from Miss K. B., an outgoing twenty-two-year-old Ketchikan retail clerk, interviewed in 2000.

"My mother told me several years ago that around 1980 she and my father were sport-trolling for salmon in Mink Bay, in the middle of the peninsula that forms the southern tip of the Alaskan mainland Misty Fjords National Monument. Dad was at the wheel of the boat and looking ahead, but my mother was watching the shore, which was only about fifty feet away. Mom said that for just a few seconds, she saw the bushes parted by black, hairy arms and hands, and she watched a tall, heavily built, manlike creature staring at the boat. She said it was definitely not a bear and had a face like more like a man. What she insisted was that, for several seconds, she and the creature 'made eye contact.' She said that Dad didn't see it. He was looking the other way. I believe she was telling me exactly what she had seen." [S81A-map3]

There is one report from the southern Alaskan Mainland that did not occur on either a road, nor along a shore, but is included here simply because it needed telling. It is the story of three snowmobile enthusiasts who reported encountering a sasquatch one winter along the U.S./Canada border above Hyder, Alaska, and Stewart, B.C. Outdoorsman Frank S., a Ketchikan contractor, related in an interview in 2000 how he and two friends had a brush with a sasquatch, which they also at first mistook for a bear, on a snowfield above Hyder in the winter of 1997.

"In February, '97 my friends and I had taken snow machines up Hidden Glacier north of Hyder, up the Long Lake Road. We were running the machines at night up above the tree line and it was about 3:00 a.m. We had reached a point where the slope was too steep to go on, and we had turned the machines around and shut down for a rest. There was good moonlight up on the snow. You could see miles.

"We all noticed what we at first thought was a man-shaped rock about 200 feet away. Turning the headlamps on it we saw it stand up, a tall animal with a whitish eye-shine just standing there. My first thought, although it was winter, was that it was a bear. But when it screamed and started walking toward us, we changed our minds! Its arms moved with a swing and it was big—maybe 800 to 1,000 pounds. We stayed on our machines watching it for a total of about two minutes but when it started walking towards us, we turned our machines quick and went back down to the trucks." [S97B-map6]

On the surface, this story is not that different from many of the other reports, but on second glance it is about as unusual as they get. The time of year falls into the "seldom-seen" period, one of only two February reports, and whereas the Kake sighting was also in snow, that smaller creature apparently had the common sense or instinct to winter on the coast and avoid man and his machines. In the Hyder report, not only did the creature reportedly advance after watching, but also it was heard uttering a scream. While there is no proof that sasquatches hibernate, it is interesting to speculate if the threatening approach and scream were perhaps simply the bad attitude any hibernator could be expected to have on being awakened from a sound sleep by two-cycle engines directly above or outside a cozy winter abode. This is exactly the type of vocalization that some people have occasionally reported when they have pulled noisily up in a truck, car or boat at some remote, and hitherto peaceful, wilderness setting. This subject of ritual threatening behavior will be explored in greater detail further on, but given the time of the sighting in the Hyder snowfield case, the sasquatch appears far more likely to have simply been in transit between two main watersheds and not aroused from any wintertime siesta.

1. Krantz, G., 1999, *Bigfoot Sasquatch Evidence*, Hancock House Publishers, Surrey, B.C., p 78.

7 Chasing Deer and Other Stories

Al Jackson, a sixty-five-year-old retired mill worker now living in Klawock, Prince of Wales Island, has researched dozens of sasquatch tales in southeast Alaska. A compact Tlingit man, friendly and energetic, Al is recognized and greeted with a smile almost everywhere he goes on Prince of Wales. He has an instinct for detective work and enjoys getting to the bottom of things. When he forwards sasquatch reports or older sasquatch tales in person, his eyebrows seem to form a permanent arch over twinkling eyes. In just such a way in 1996, he passed on to me the following second-hand story from the rugged mountains of Misty Fjords National Monument on the southern tip of Alaska.

"'Old Man S.' a well-known Ketchikan lumberman, before he died three winters ago, told me in 1993 that he had been mountain goat hunting in 1955 with Don B. and two other men, both loggers, up above Rudyerd Bay in Misty Fjords. They went up the mountain above their boat and they walked up this draw, still in some trees, and he, 'Old Man S.,' said, 'The damnedest thing happened. We were just about to break out of the timber, you know, to get into the meadows, and right above us twenty-five to thirty goats come stampeding down on us. We each shot one, and the other guys all they wanted was the hide. I skinned mine out and was working on the hindquarters meat and 'backstrap,' the choice back cuts of meat, and the guy who was up in front, about a hundred feet ahead of me, had skinned his hide and he was watching me butcher the goat. While I was working, he said to me, 'Did you see that thing that was running those goats up there?' I said, 'No,' and he said to me, 'You should have seen it! It was about ten

Figure 19: Artist's impression of a sasquatch reported driving goats above Rudyerd Bay, East Behm Canal, Alaska, in 1955.

feet tall and all covered with black hair, chasing those goats, and it took off when we all started shooting.'

"Old man S. continued, 'I didn't say nothing, but you know, later on, on the boat having coffee on deck, he said to me again, "Man! You should have seen that thing! Ten feet tall, hair all over and long skinny legs!"' [S55A-map3]

Jackson summed up with a recollection that the old man mentioned the group finding large manlike tracks, about nineteen inches long, while up the mountain. (Curiously, this is almost the same exact spot where in 1998, more than forty years later, Ms. S. L. and her companion heard a series of five loud, angry-sounding screams.)

Jackson also noted that there had been one other Revilla' Island sighting in 1955, involving a relative of his. I have since been able to verify, from one of the witnesses, the original report just as it is stated here by Mr. Jackson. It reads: "My cousin Will K., who has since passed on, told me a few years ago that when he was a boy of thirteen, in 1955, he was walking with four other kids on North Tongass Road at Cranberry Road, about twelve miles north of Ketchikan. He told me, 'It was a sunny summer's day, around noon. Suddenly we saw a deer come crashing out of the woods on the bluff above us, and right behind it came a great big black thing. Both the deer and the creature passed within thirty feet of the nearest boy and it appeared to me to be about ten feet tall. The deer crossed the road at top speed but the thing saw us, seemed to stop for a second, and then took off right after the deer again. It was real big and ran on two legs, covered all over with black hair. Later my stepfather, W. K. Sr., said to me, 'Don't tell anyone, because someone might want to shoot it." [S55C-map5]

The North Tongass Highway location given is almost exactly the same as in two other sighting reports, those of Mrs. K. W. in the late 1970s [S70B] and cyclist Gerald P in 1998 [S98A]. Whipple Creek flows through this stretch of highway, and only a mile upstream, one track report [T98A] and one possible vocalization report have been noted. Blacktail deer abound in the area, and would almost certainly form part of the diet of any sasquatches on this southwest corner of Revilla' Island.

Deer are everywhere in "southeast," but what about the other large mammals for which Alaska is known? One question sometimes posed is, if sasquatches exist in southeast Alaska, especially in brown bear habitat, how could the two species coexist? One hypothetical answer may be contained in the following report from Thomas J. Fisher, an experienced Ketchikan commercial fishing skipper and hunter. Fisher is a straightforward man, not afraid to speak out. He punctuates his words with a thrust of his close-bearded jaw and a jab of his finger. In several repeated interviews in 2000, he related an unusual sasquatch/ brown bear sighting back in 1985. He stated that in July of that year he had been on his commercial salmon troller off the northern tip of Yakobi Island, which is just west of the north end of Chichagof Island, north of Sitka and west of Juneau.

"One evening in July, we had just anchored on the north end of Yakobi Island, in Soapstone Bay, about 100 yards offshore, and we were just starting to eat supper. It was a beautiful evening, about seven or eight, with a clear sky. My crewman and I were sitting on deck watching a huge brown bear in the beach grass on the left side of the bay, about a hundred yards away. It was big. Its belly was above the tops of the grass and I would estimate the bear would have squared about thirteen feet. The bear didn't seem bothered by our boat or the Hoonah fish packer boat near us at all. All of a sudden, we saw the bear's hair go up and it took off to the left, looking back like it was frightened by something over to the right on the beach.

"We couldn't figure what had scared it. About fifteen minutes later, we both noticed something tall and black moving on the beach about 300 yards to the right. It was shaped like a huge, hair-covered man with long black hair all over. It stopped in the tall grass to look at us. We watched it staring at us for about fifteen seconds. The grass just came up to its knees. The crewman and I looked at each other and we both said to each other, 'Are we looking at what I think we're looking at?'

"It seemed to be fully erect like a man, and we estimated its height at about nine feet. Then it took off running to the right in the opposite direction from the bear, and covered about 100 yards faster than anything on two legs I've ever seen. It definitely wasn't human, but it didn't look like an ape in its features. It ran just like a man, but faster and it was heavier built. I've hunted bears, and there was no way it was a bear!

"The next day we walked along through the tall grass all along the shore. It was all about three feet high. Two good-sized black-tail bucks we saw standing in the grass were covered by the grass to above their shoulder height. It was clearly a bigfoot we saw. I don't worry about what people may say, what I've seen is good enough for me." [S85A-map1]

Thomas Fisher sticks by his story. Were the two reported animals startled by the scent of man? Perhaps; but the respective directions in which they were reported to have fled suggested otherwise.

The forest at Soapstone Bay is mature hemlock and spruce, thick and prolific. Recently it has been set aside as a conservation area. Huckleberry, salal and devil's club grow where light filters down. The shoreline is abundant with marine life and there is no community on Yakobi Island. If, as Mr. Fisher attests, there are sasquatches on Yakobi, then there could well be sasquatches everywhere in southeast Alaska. Given the nature of the forest and coast it would be difficult to dispute.

Not everyone would welcome having sasquatches everywhere in southeast Alaska. In a well-written and illustrated Alaskan natural history, *The Nature of Southeast Alaska*, by Rita M. O'Clair, Robert H. Armstrong and Richard Carstensen, all accomplished and respected biologists or teachers, there exists a good example of how unwilling established science may be to embrace such a species as the sasquatch. In referring to the popularity of bears the authors write, "When conversation lags, we revive it with bears. There's no need for Sasquatch in Southeast Alaska, we have Brown Bear."[1] Of course, the authors may be speaking simply without bias from an ecological perspective, merely commenting that on the "brown bear islands" of southeast Alaska (Admiralty, Baranof and Chichagof), the niche that might provide a suitable lifestyle for a sasquatch is already occupied by brown bears. (The bears do not occur on the other islands of "southeast," with rare exceptions crossing from the mainland to Revilla' Island.) But if bears don't run down adult deer or forage frequently in the ocean, as has been suggested for sasquatches in southeast Alaska, then perhaps a nocturnal year-round omnivore such as the sasquatch might not find it difficult to coexist there, even when bears are not hibernating.

1. O'Clair, R. M., R. H. Armstrong and R. Carstensen, 1992, *The Nature of Southeast Alaska*, Alaska Northwest Books, Anchorage & Seattle, p. 72.

8   A Sightings Digest

Because the human population of southeast Alaska is so sparsely distributed, and in most cases confined to isolated villages to which access is difficult, the gathering and investigation of sasquatch reports are likely much more difficult than in the continental United States. Local reports often go unmentioned to anyone but family and perhaps a few friends, and almost always go unnoticed and undocumented. However, the preceding sixty-odd sighting reports, mostly from southern southeast Alaska and spanning just more than fifty years, are fairly uniform in the details of the creatures described. It is not likely that many of them are describing the same individual animals, but one cannot tell simply from an interview. It is very doubtful, given the locals' solid familiarity with bears, that many would be confusing what they saw with a bear.

What does appear in the reports brought to light, is the general pattern of a large, hair-covered hominid, indistinguishable in any way from reports of sasquatches documented in the Pacific Northwest and western Canada. Their height, where reported, varies from just under man-sized to several at ten feet, with a modal average somewhere just over seven feet. The only report of a clearly identified female detailed seven feet for an estimate of stature. Most all reports mentioned a heavy build, with usually no mention made of a neck. When it was mentioned, the neck was often described as "thick." Heavy shoulders were mentioned several times. The hair color of sasquatches reported in southeast Alaska would appear to be similar to the variety of coloration described elsewhere. Although sasquatch colors in southeast Alaska have been most often reported as "dark," i.e., black or dark brown, other colors such as reddish brown, auburn, chestnut and beige are also reported. There are only two mentions of grayish fur and none of silver or white, although this could change as reports accumulate.

In most cases the animals reported seemed shy and retiring, not stopping for more than a moment to investigate human presence, but in a few cases more aggressive behavior has been mentioned. Where vehicles penetrating into remote areas have been involved, the nature of the reported behavior tended toward more aggressive displays. In the preceding reports and also in a following chapter dealing with purported sasquatch vocalizations, it would seem that driving into a remote location and parking a vehicle is more likely to be a factor in precipitating aggressive behavior. With reports nearer centers of population, parked vehicles do not seem to have been related to aggressive responses. Anchoring a boat in a remote setting has been noted in four instances that have apparently elicited screams or yells or, in one case, thrown objects.

Reports of sasquatches seemingly ignoring human presence are noted in two reports involving a sasquatch apparently digging on a beach, and in six reports where the motorists commented particularly on how the creature failed to look in the direction of the vehicle. Although bears, for example, often look into oncoming headlight beams, any sasquatch having done so previously might have learned to look away. There is certainly the suggestion of some degree of heightened intelligence. Five particular reports on land mention approaching or curiosity behavior. Three of these seemed peaceful. In the one case of Ms. Shannon H. of Ketchikan, whose group was followed into a Ward Lake picnic shelter at night, there was some evidence of curiosity. In J. W. Huff's Bradfield Canal report of 1969, the sasquatch spent a good deal of time simply watching the two men. And in the startling 1979 report of fishing captain Harold A. of Sitka, the sasquatch apparently had an interest in the skipper himself as he reposed beside a galley porthole at night on his boat.

There are at least three other reports describing a sasquatch directly approaching people. Mr. Larry Thomas (1992 report near Ward Lake) recounted an upright creature approaching him for a few steps after he whistled at it. In another, that of Mr. Frank S. (with others on snowmobiles above Hyder in 1992), there was an apparent belligerence on the part of the sasquatch as it approached the men. And the six and a half foot *Zinjanthropus*-looking creature reported near Juneau in 2001 was apparently quite deliberate in approaching three individuals seated around a campfire. All of these various behaviors are not much different from the general behavior of the great apes toward man, that is primarily retiring, but also on occasion aggressive, curious or even, apparently, playful.

What all of these reports seem to suggest is that sasquatches, if not shot at or otherwise molested, are actually quite benign, calm-mannered and even curious creatures. There was some allusion to attempted violence at times by the creatures reported around the house at Klawock, but to my knowledge, there are few or no accounts of anyone ever being attacked or hurt. In fact it is worth noting that in all of the southeast Alaskan reports collected here, there have only been two accounts alleging violence toward man. Both are old, second- and third-hand, and only in one was there mention of even minor human injury. (See Watson's and Colp's stories in chapter 9.) I have discounted "missing-person" stories, even when there are locations that seem to have more than their share of such unfortunate mishaps. Granite Basin, a closed watershed north of Ketchikan has a supposed history of four individuals never found, but Alaska is full of such places. It is a tough land where surviving relatives sometimes never get closure on a missing family member. It can hardly be fair to blame it all on sasquatches.

What we are left with so far is the description of an erect, hirsute creature that is nocturnal, apparently swims quite well, is probably capable of taking large game animals by ambush or persistent running and exhibits a routine curiosity toward man and his tools. Although, in southeast Alaska, the old Klawock stories and one track report suggest sasquatches may move in small family units or form loose associations, it would appear from most reports that, in Alaska at least, solitary behavior is the norm where a sasquatch is seen. Most reported sasquatches are apparently males, and if they are roaming adolescents, perhaps family groups are simply remaining in more remote areas, unobserved. It is apparent as well that sasquatches are less often seen in daytime and exhibit at least the modicum of intelligence required to avoid man, if they so choose.

Having a coat that is usually dark in color, and being for the most part nocturnal, it would be rather easy for a sasquatch to stand quite close to a forest trail or road without being observed. One may wonder just how many experiences of "feeling watched" or "having the hair go up" has involved the near presence of an unobserved sasquatch. This phenomenon is so widespread and often reported in Alaska and elsewhere that it could fill a chapter of reports, but for obvious reasons, many of them subjective, they are not included here. Many individuals participating in personal field investigation of sasquatch reports spend a lifetime without actually observing a creature, and yet one often hears of people in the woods who at times had the sensation of being watched. Perhaps the subliminal pheromones of body chemistry are responsible. This shouldn't be taken to imply that behind every such report there is a sasquatch lurking. Then again, who could tell?

9 Old Stories from the B.C./Alaska Coast

The coasts of British Columbia and southeastern Alaska are huge. Between Yakutat, Alaska, and the lower mainland of British Columbia, there are thousands of miles of coastline, almost entirely unpopulated. Not much is ever written about it. Nevertheless, from time to time on these tangled shores, there have surfaced stories of wild, hair-covered hominids, generally matching the description of sasquatches. A few of the older stories have been published, many have not. The earliest written accounts of sasquatches by non-Natives on western shores are a bridge between Native beliefs and a larger body of more recent reports. It is to these early stories we should now turn for a first look at what has now become a favorite subject of popular media and increasing scientific scrutiny—the sasquatch of the Pacific Northwest. The following are a collection of accounts stated to have taken place in the first half of the twentieth century, of creatures which may collectively be considered as sasquatches.

Clayton Mack and Sasquatches

One of the most colorful and well liked of all British Columbia grizzly guides was a Bella Coola Native, the late Clayton Mack. In an engaging narrative of his life and times, Bella Coola family physician and friend of Mack, Dr. Harvey Thommassen, recorded a good number of his outdoor and guiding experiences. These are faithfully reproduced in Mack's earthy vernacular in a recently published book, *Grizzlies and White Guys*.[1] In a chapter entitled "The Sasquatch," Mack relates three accounts of encounters with sasquatches, excerpts of which are noted as follows (by kind permission of Harbour Publishing).

"I was probably in my thirties [the 1940s]. I had a little boat, about a thirty-foot boat with a single cylinder engine. I got to Jacobsen Bay, about fifteen miles from Bella Coola when I saw something right out on low tide. I saw something on the edge of the water. It was kneeling down, like, and I could see his back humping up on the beach. It looked like he was lifting up rocks or maybe digging clams. But there was no clams there. I turned the boat right in toward him, I wanted to find out what it was. For a while I thought it was a grizzly bear, kind of a light color fur on the back of his neck, like a light brown, almost buckskin color, fur. I nosed in toward him to almost seventy-five yards to get a good look.

"He stood straight up on his hind feet, straight up like a man, and I looked at it. He was looking at me. Gee, It don't look like a bear, it had arms like a human being, it has legs like a human being and it got a head like us. I keep on going in toward him. He started to walk away from me, walking like a man on two legs. He was about eight feet high. He got to some drift logs, stopped and looked back at me. Looked over his shoulder to see me. Grizzly bear don't do that, I never see a grizzly bear run on its hind legs like that and I never see a grizzly bear look over its shoulder like that. I was right close to the beach now. He stepped up on those drift logs, and walked into the timber. Stepped on them logs like a man does. The area had been logged before, so the alder trees were short, about eight to ten feet high. I could see the tops moving as he was spreading them apart to go through. I watched as he went a little higher up the hill. The wind blew me in toward the beach, so I backed up the boat and keep on going to Kwatna Bay."[2]

Mack goes on to relate five more sasquatch encounters of various sorts, one of which involved an American hunter from northern California. This man had accompanied Mack on a hunting trip to the Bella Coola area around 1973, up to the Asseek River at the extreme head of South Bentick Arm. One evening, the men were walking the beach flats back to the boat when they sighted what Mack at first thought was a black bear a ways down the beach.

"When I got closer, not too far now, the hunter grabbed the collar of my shirt and pulled me back.

"'Clayton,' he said, 'That's not a black bear, that's a sasquatch.' He keep on saying, 'It's a sasquatch.'

"I didn't say nothing, I started walking again. I said, 'Stay right behind me.' He was only about seventy-five yards away.

"'Clayton,' he say again, 'that's not a black bear, that's a sasquatch.'

"I kneel down on the ground, I turned toward him, 'What do you know about sasquatch?'

"He says, 'I come from North California. We get them in that country. In the big mountains that get snow on them. Those mountains in Northern California which have glaciers on them. Some people hunt them,' he said.

"I said, 'How do they look like?'

"He said, 'Well, you seeing one there now, that's how they look like!'

"And I started walking again. I get pretty close now. Then that 'black bear' stand up on his both legs, and he look at me. I keep going closer. Gee, I was pretty close now. He started looking at me, making no noise or anything. I feel the barrel of a gun against my cheek. I pushed that hunter's gun away from my face. 'Don't shoot him,' I said.

"That hunter whispered in my ear, 'Look through your scope and see how he look like.'

"I turned the scope to 4 power—wide and close—four times closer than naked eye. I looked through that scope, I look at his mouth, look like rice. Little white thing in his mouth, look like rice. I look at his lips kind of turnin' in and turnin' out, the top and the bottom too. I look at his face and his chest. The shape of his face is different than a human being face. Hair over face, eyes were like us, but small. Ears small too. Nose just like us, little bit flatter, that's all. Head kind of look small compared to body. Looks friendly, doesn't look like he's mad or has anything against us. Didn't snort or make sound like a grizzly bear. On the middle of his chest, looked like a line of no hair, hair split apart a bit in the middle. Skin is black where that hair split apart. It was a male, I think. I can't—no way I can—shoot him. I had a big gun too. Big gun, .308. I aim, had my finger on the trigger, point it right at the heart. One shot kill him dead, just like that. I couldn't shoot him. Like if a person stand over there, I shoot him, same thing. No way I can kill him.

"My mother told me, 'Don't ever kill sasquatch, don't shoot 'em. If you shoot 'em, you gonna lose your wife, or else your mother or your dad or else your brother or sister. It will give you bad luck if you shoot them, kill them. Leave them alone,' she said. If you see one, walk the other way, let them walk that-a-way.' That's why I don't want to shoot one. My mother had seen them. She hear them too. A lot of Indian people seen them in the old days."[3]

Mack concluded that on one particular occasion he had heard one of their calls, which sounded to him like "Haii Haii Haii," and that they were well known to his Bella Coola people as "*boqs*."

All of Clayton Mack's stories were recorded on tape by Dr. Thommassen in the vernacular of Mack's time and place. Mack was a respected man in his community and regarded as being as good as his word. His descriptions of sasquatches are so straightforward that one really is compelled to accept them or reject them outright. Although these and other accounts may seem unusual or startling to many, his collected stories reinforce the genuine nature of Clayton Mack, who since his passing in 1993 is still remembered and highly regarded in Bella Coola and the surrounding country. Finally, they provide a fascinating background to anyone who might wish to further examine the Bella Coola area, a wild and spectacular region of the West Coast.

The Sasquatch in the Glacier

One of the oldest accounts of sasquatches from the Alaska/B.C. coast is contained in the story of Frank E. Howard[4], in a 1908 edition of the *Alaska-Yukon Magazine*. It was said to have taken place one July, some years previously, at southeastern Alaska's northernmost bay. In the story, Howard recounted how, on a prospecting trip north of Yakutat some time previously, he had become lost on one of the arms of the Malaspina Glacier. He had followed the edge of the glacier by canoe and secured his craft in a sheltered spot where the glacier met the beach. Attempting to cross it on crampons to a barren rocky ridge that he had seen from the head of Yakutat Bay, he had the misfortune to fall into a narrow crevasse. Determining he could not climb back up, he resolved to find a way down through the crevasses to the edge of the glacier.

"I arose and started down the slope [inside the crevasse] with the idea of reaching the water and following along its margin while the tide was low, in search of some crevasse leading out into the open bay. I was sure the great cavern was crevassed to the surface at some point beyond.

"As I kept going ahead I noticed a gradual increase of light, and in a few more steps, I stood in a broad wall of blue light that came down from above and, looking up, I saw there was no clear opening to the surface. But objects were now revealed some distance around." Just as Howard began to see more clearly in the faint light of the ice cavern, a movement caught his attention.

"Then an object rose slowly out of the glimmer and took form— a spectral thing, with giant form, and lifelike movement. The object rose erect, a goliath in the shape of a man. Then, watching me with a slantwise glance, it walked obliquely from me, until its form faded in the gloom of the cavern. In a concluding remark on the manlike creature, he added, "that with its shaggy light-colored fur and huge size, the creature in some ways resembled a bear with bluish gray fur, but that it had a roughly human form, and at all times walked erect." [OS09-map1]

Howard then recounted his passage out by following a watercourse through the crevasses out to a timbered shore and from there to his camp. It seems plausible that this event may have happened, in some form or another, as described. Howard's account has also been edited and abridged by a Juneau author, Ed Ferrell, in a book, *Strange Stories of Alaska and the Yukon*,[5] with a footnote: "There is no further record of Frank Howard. Yet his account of the creature he encountered in the glacier is consistent with other reports of Big Foot or Sasquatch." At any rate it is an intriguing tale, if not simply a good yarn, and Ferrell's summation of the creature described as "consistent with sasquatch" seems quite appropriate.

The Albert Ostman Story—a Sasquatch Classic

The most detailed account of an encounter with sasquatches anywhere is certainly that of the late Albert Ostman. A Swedish-Canadian outdoorsman, Ostman quietly claimed to have to have tangled up with a family of sasquatches while penetrating the vast mountains of coastal British Columbia on a prospecting trip near Toba Inlet, B.C., in 1924. In the 1950s, after he had retired to the Fraser Valley, he was variously interviewed on different occasions by Ivan Sanderson, John Green and Rene Dahinden, newspapermen and even a magistrate. His story never faltered, although some additional questioning by Green added a few details. Some of the interviews were protracted and rather grueling, intended to trip him up, but in the different interviews Ostman stuck by his story with patience and steady demeanor. When interviewed by Green, Ostman's account was already written and this enabled Green to verify previous statements and elucidate several details not included in some of the other interviews. When Ostman was later asked to swear to the accuracy of his account, some thirty years after the event, he made clear he could do so only as to the main elements of it and not to every small detail.

Mr. Ostman's fascinating story, recorded in interviews in 1957, is really worth reading in its entirety and is here reproduced verbatim (by kind permission of John Green).

"I have always followed logging and construction work. This time I had worked for over one year on a construction job and thought a good vacation was in order. B.C. is famous for lost gold mines. One is supposed to be at the head of Toba Inlet,—why not look for this mine and have a vacation at the same time? I took the Union Steamship boat to Lund, B.C. From there I hired an old Indian to take me to the head of Toba Inlet.

"This old Indian was a very talkative old gentleman. He told me stories about gold brought out by a white man from this lost mine. This white man was a very heavy drinker—spent most of his money freely in saloons. But he had no trouble getting more money. He would be away a few days, then come back with a bag of gold. But one time he went to his mine and never came back. Some people say a Sasquatch had killed him.

"At that time I had never heard of Sasquatch. So I asked what kind of animal he called a Sasquatch. The Indian said, 'They have hair all over their bodies, but they are not animals. They are people. Big people living in the mountains. My uncle saw the tracks of one that were two feet long. One old Indian saw one over eight feet tall.'

"I told the Indian I didn't believe in their old fables about mountain giants. It might have been some thousands of years ago, but not nowadays.

"The Indian said, 'There may not be many, but they still exist.'

"We arrived at the head of the inlet at 4:00 p.m. I made camp at the mouth of a creek. The Indian was in no hurry, he had to wait for the high tide to go back. That would be about 7:00 p.m. I tried to catch some trout in the creek, but no luck. The Indian had supper with me, and I told him to look out for me in about three weeks. I would be camping at the same spot when I came back. He promised to tell his friend to look out for me too.

"I spent most of the forenoon looking for a trail but found none, except for a hog-back running down to within about a hundred feet of the beach. So I swamped out a trail from there, got back to my camp about 3:00 p.m. that afternoon and made up my pack to be ready in the morning. My equipment consisted of one 30-30 Winchester rifle, I had a special home-made prospecting pick, axe on one end, and pick on the other. I had a leather case for this pick which fastened to my belt, also my sheath knife.

"Next morning I took my rifle with me but left my equipment at the camp. I decided to look around for some deer trail to lead me up into the mountains. On the way up the inlet I had seen a pass in the mountainside that I wanted to go through, to see what was on the other side.

"The storekeeper at Lund was co-operative. He gave me some cans for my sugar, salt and matches to keep them dry. My grub consisted mostly of canned stuff, except for a side of bacon, a bag of beans, four pounds of prunes and six packets of macaroni, three pounds of pancake flour, cheese, and six packets Rye King hard tack, three rolls of snuff, one quart sealer of butter and two one-pound cans of milk. I had two boxes of shells for my rifle.

"The storekeeper gave me a biscuit tin, I put a few things in it and cached it under a windfall, so I would have it when I came back here waiting for a boat to bring me out. My sleeping bag I rolled up and tied on top of my pack sack—together with my ground sheet, small frying pan, and one aluminum pot that held about a gallon. As my canned food was used, I would get plenty of empty cans to cook with.

"The following morning I had an early breakfast, made up my pack, and started up this hog-back. My pack must have been at least eighty pounds, besides my rifle. After one hour, I had to rest. I kept resting and climbing all that morning. About 2:00 p.m. I came to a flat place below a rock bluff. There was a bunch of willow in one place. I made a wooden spade and started digging for water. About a foot down I got seepings of water, so I decided to camp here for the night, and scout around for the best way to get on from here.

"I must have been up near a thousand feet. There was a most beautiful view over the islands and the strait—tugboats with log booms, and fishing boats going in all directions. A lovely spot. I spent the following day prospecting around. But no sign of minerals. I found a deer trail going towards this pass that I had seen on my way up the inlet.

"The following morning I started out early, while it was cool. It was steep climbing with my heavy pack. After a three hours climb, I was tired and stopped to rest. On the other side of a ravine from where I was resting was a yellow spot below some small trees. I moved over there and started digging for water.

"I found a small spring and made a small trough from cedar bark and got a small amount of water, had my lunch and rested here till evening. This was not a good camping site, and I wanted to get over the pass. I saved all the water I got from this spring, as I might not find water on the other side of this pass. However, I made it over the pass late that night.

"Now I had downhill and good going, but I was hungry and tired, so I camped at the first bunch of trees I came to. I had about a gallon of water so I was good for one day. Of course, I could see rough country ahead of me, and was trying to size up the terrain—what direction I would take from here. Towards the west would lead to low land and some other inlet, so I decided to go in a northeast direction, but I had to find a good way to get down.

"I left my pack and went east along a ledge but came to an abrupt end—was two or three hundred feet straight down. I came back, found a place only about fifty feet down to a ledge that looked like good going for as far as I could see. I got down on this ledge all right and had good going and slight down hill all day—I must have made ten miles when I came to a small spring and a big black hemlock tree.

"This was a lovely campsite. I spent two days here just resting and prospecting. There were some minerals but nothing interesting. The first night here I shot a small deer (buck) so I had plenty of good meat, and good water. The weather was very hot in the daytime, so I was in no hurry, as I had plenty of meat.

"When I finally left this camp, I got into plenty of trouble. First I got into a box canyon, and had to come back almost where I started this morning, when I found a deer trail down to another ledge, and had about two miles of good going. Then I came to another canyon, and on the other side was a yellow patch of grass that meant water. I made it down into this canyon, and up on the other side, but it was tough climbing. I was tired and when I finally got there I dug a pit for water and got plenty for my needs. I only stayed here one night, it was not a good camping site. Next day I had hard going. I made it over a well timbered ridge into another canyon. This canyon was not so steep on the west side, but the east side was almost plumb. I would have to go downhill to find a way out. I was now well below the timber line.

"I found a fair campsite that night, but moved on the next morning. It was a very hot day, not a breath of wind.

"Late that day I found an exceptionally good campsite. It was two good-sized cypress trees growing close together and near a rock wall with a nice spring just below those trees. I intended to make this my permanent camp. I cut lots of brush for my bed between these trees. I rigged up a pole from this rock wall to hang my pack sack on, and I arranged some flat rocks for my fireplace for cooking. I had a real classy setup. I shot a grouse just before I came to this place. Too late to roast that tonight—I would do that tomorrow.

"And that's when things began to happen. I am a heavy sleeper, not much disturbs me after I go to sleep, especially on a good bed like I had now.

"Next morning I noticed things had been disturbed during the night. But nothing missing that I could see. I roasted my grouse on a stick for breakfast—about 9:00 a.m. I started out prospecting. I always carried my rifle with me. Your rifle is the most important part of your equipment.

"I started out in a southwest direction below the way I had come in the night before. There were some signs (minerals) but nothing important. I shot a squirrel in the afternoon, and got back to camp at 7:00 p.m. I fried the squirrel on a stick, opened a can of peas and carrots for my supper and gathered up dry branches from trees. There are always dead branches of fir and hemlock under trees, near the ground. They make good fuel and good heat.

"That night I filled up the magazine of my rifle. I still had one full box of 20 shells in my pack, besides a full magazine and six shells in my coat pocket. That night I laid my rifle under the edge of my sleeping bag. I thought a porcupine had visited me the night before and porkies like leather, so I put my shoes in the bottom of my sleeping bag.

"Next morning my pack sack had been emptied out. Someone had turned the pack upside down. It was still hanging on the pole from the shoulder straps I had hung it up. Then I noticed one half-pound package of prunes was missing. Also my pancake flour was missing, but my salt bag was not touched. Porkies always look for salt, so I decided it must be something else than porkies. I looked for tracks but found none. I did not think it was a bear, they always tear up and make a mess of things. I kept close to camp those days in case this visitor would come back.

"I climbed up on a big rock where I had a good view of camp, but nothing showed up. I was hoping it would be a porky so I could get a good porky stew. These visits had now been going on for three nights.

"I intended to make a new campsite the following day, but I hated to leave this place. I had fixed it up so nicely, and these two cypress trees were bushy. It would have to be a heavy rain before I would get wet, and I had good spring water that is hard to find.

"This night it was cloudy and looked like it might rain. I took special notice of how everything was arranged. I closed my pack sack, I did not undress, I only took off my shoes, put them in the bottom of my sleeping bag. I drove my prospecting pick into one of the cypress trees so I could reach it from my bed. I also put my rifle alongside me, inside my sleeping bag. I fully intended to stay awake all night to find out who my visitor was, but I must have fallen asleep.

"I was awakened by something picking me up. I was half asleep and at first I did not remember where I was. As I began to get my wits together, I remembered I was on this prospecting trip, and in my sleeping bag.

"My first thought was—it must be a snow slide, but there was no snow around my camp. Then it felt like I was tossed on horseback, but I could feel whoever it was, was walking.

"I tried to reason out what kind of animal this could be, I tried to get at my sheath knife, and cut my way out, but I was in an almost sitting position, and the knife was under me. I could not get hold of it, but the rifle was in front of me, I had a good hold of that, and had no intention to let go of it. At times I could feel my pack sack touching me, and could feel the cans in the sack touching my back.

"After what seemed like an hour, I could feel we were going up a steep hill. I could feel myself rise for every step. What was carrying me was breathing hard and sometimes gave a slight cough. Now, I knew this must be one of the mountain Sasquatch giants the Indian told me about.

Figure 20: Author's impression of a family of sasquatches, as reported by Albert Ostman, alleged abductee, near Toba Inlet, B.C., in 1924. [S24A-map11]

"I was in a very uncomfortable position, unable to move. I was sitting on my feet, and one of the boots in the bottom of the bag was crossways with the hobnail sole up across my foot. It hurt me terribly but I could not move.

"It was very hot inside. It was lucky for me this fellow's hand was not big enough to close up the whole bag when he picked me up—there was a small opening at the top. Otherwise I would have choked to death.

"Now he was going downhill, I could feel myself touching the ground at times and at one time he dragged me behind him and I could feel he was below me. Then he seemed to get on level ground and was going at a trot for a long time. By this time, I had cramps in my legs, the pain was terrible. I was wishing he would get to his destination soon. I could not stand this type of transportation much longer.

"Now he was going uphill again. It did not hurt me so bad. I tried to estimate distance and directions. As near as I could guess we were about three hours traveling. I had no idea when he started as I was asleep when he picked me up.

"Finally he stopped and let me down. Then he dropped my pack sack, I could hear the cans rattle. Then I heard chatter—some kind of talk I did not understand. The ground was sloping so when he let go of my sleeping bag, I rolled over head first downhill. I got my head out, and got some air. I tried to straighten my legs and crawl out but my legs were numb.

"It was still dark, I could not see what my captors looked like. I tried to massage my legs to get some life in them, and get my shoes on. I could hear now and it was at least four of them. They were standing around me and continuously chattering. I had never heard of Sasquatch before the Indian told me about them. But I knew I was right among them. But how to get away from them, that was another question. I got to see the outline of them now as it began to get lighter.

"I now had circulation in my legs, but my left foot was very sore on top where it had been resting on my hobnail boots. I got my boots out from the sleeping bag and tried to stand up. I was wobbly on my feet but had a good hold of my rifle.

"I asked, 'What you fellows want with me?'

"Only some more chatter.

"It was getting lighter now, and I could see them quite clearly. I could make out forms of four people. Two big ones and two little ones. They were all covered with hair and no clothes on at all.

"I could now make out mountains all around me. I looked at my watch. It was 4:25 a.m. It was getting lighter now and I could see the people clearly. They looked like a family, old man, old lady and two young ones, a boy and a girl. The boy and the girl seemed to be scared of me. The old lady did not seem too pleased about what the old man dragged home. But the old man was waving his arms and telling them about what he had in mind. They all left me then.

"I had my compass and my prospecting glass on strings around my neck. The compass in my left hand shirt pocket and my glass in my right hand pocket. I tried to reason our location and where I was. I could see now that I was in a small valley or basin about eight or ten acres, surrounded by high mountains, on the southeast side there was a V-shaped opening about eight feet wide at the bottom and about twenty feet wide at the highest point— that must be the way I came in. But how will I get out?

"The old man was now sitting near this opening. I moved my belongings up close to the west wall. There were two small cypress trees there, and this will do for a shelter for the time being. Until I find out what these people want with me, and how to get away from here. I emptied out my packsack to see what I had left in the line of food. All my canned meat and vegetables were intact and I had one can of coffee. Also three small cans of milk—two packages of Rye King hard tack and my butter sealer half full of butter. But my prunes and macaroni were missing. Also my full box of shells for my rifle. I only had six shells besides what I had in the magazine of my rifle. I had my sheath knife but my prospecting pick was missing and my can of matches. I only had my safety box full and that held about a dozen matches. That did not worry me—I can always start a fire with my prospecting glass when the sun is shining, if I got dry wood. I wanted hot coffee, but I had no wood, also nothing around here that looked like wood. I had a good look over the valley from where I was—but the boy and the girl were always watching me from behind some juniper bush. I decided there must be some water around here. The ground was leaning toward the opening in the wall. There must be water at the upper end of this valley, there is green grass and moss along the bottom.

"All my utensils were left behind. I opened my coffee tin and emptied the coffee in a dishtowel and tied it with the metal strip from the can. I took my rifle and the can and went looking for water. Right at the head under a cliff there was a lovely spring that disappeared underground. I got a drink and a can full of water, when I got back the young boy was looking over my belongings, but did not touch anything. On my way back I noticed where these people were sleeping. On the east side of this valley was a shelf in the mountainside, with overhanging rock, looking something like a big undercut in a big tree about ten feet deep and thirty feet wide. The floor was covered with lots of dry moss, and they had some kind of blankets woven of narrow strips of cedar bark, packed with dry moss. They looked very practical and warm—with no need of washing.

"The first day not much happened. I had to eat my food cold. The young fellow was coming nearer me, and seemed curious about me. My one snuff box was empty so I rolled it towards him. When he saw it coming, he sprang up quick as a cat, and grabbed it. He went over to his sister and showed her. They found out how to open and close it—they spent a long time playing with it—then he trotted over to the old man and showed him. They had a long chatter.

"Next morning, I made up my mind to leave this place—if I had to shoot my way out. I could not stay much longer, I had only enough grub to last me till I got back to Toba Inlet. I did not know the direction but I would go down hill and I would come out near civilization someplace. I rolled up my sleeping bag, put that inside my pack sack—packed the few cans I had—swung the sack on my back, injected a shell in the barrel of my rifle and started for the opening in the wall. The old man got up, held up his hands as though he would push me back.

"I pointed to the opening. I wanted to go out. But he stood there pushing towards me—and said something that sounded like 'Soka soka.' I backed up to about sixty feet I did not want to be too close, I thought, if I had to shoot my way out. A 30-30 might not have much effect on this fellow, it might make him mad. I had only six shells so I decided to wait. There must be a better way than killing him, in order to get out from here. I went back to my campsite to figure out some other way to get out.

"If I could make friends with the young fellow or the girl, they might help me. Then I thought of a fellow who saved himself from a mad bull by blinding him with snuff in his eyes. But how will I get near enough of this fellow to put the snuff in his eyes? So I decided next time I give the young fellow my snuff box to leave a few grains of snuff in it. He might give the old man a taste of it.

"But the question is, in what direction will I go, if I should get out? I must have been near 25 miles northeast of Toba Inlet when I was kidnapped. This fellow must have traveled at least 25 miles in the three hours he carried me. If he went west we would be near salt water—same thing if he went south—therefore he must have gone northeast. If I then keep going south and over two mountains, I must hit salt water someplace between Lund and Vancouver.

"The following day I did not see the old lady until about 4:00 p.m. She came home with her arms full of grass and twigs of all kinds from spruce and hemlock as well as some kind of nuts that grow in the ground. I have seen lots of them on Vancouver Island. The young fellow went up the mountain to the east every day, he could climb better than a mountain goat. He picked some kind of grass with long sweet roots. He gave me some one day—they tasted very sweet. I gave him another snuff box with about a teaspoon of snuff in it. He tasted it, then went to the old man—he licked it with his tongue. They had a long chat. I made a dipper from a milk can. I made many dippers—you cut two slits near the top of any can—then cut a limb from any small tree—cut down back of the limb—down the stem of the tree—then taper the part you cut from the stem. Then cut a hole in the tapered part, slide the tapered part into the slit you made in the can, and you have a good handle on your can. I threw one over to the young fellow that was playing near my camp, he picked it up and looked at it then he went to the old man and showed it to him. They had a long chatter. Then he came to me, pointed at the dipper then at his sister. I could see that he wanted one for her too. I had other peas and carrots, so I made one for his sister. He was standing only eight feet away from me. When I had made the dipper, I dipped it in water and drank from it, he was very pleased, almost smiled at me. Then I took a chew of snuff, smacked my lips, said, 'that's good.'

"The young fellow pointed to the old man, said something that sounded like 'Oook." I got he idea that the old man liked snuff, and the young fellow wanted a box for the old man. I shook my head. I motioned with my hands for the old man to come to me. I do not think the young fellow understood what I meant. He went to his sister and gave her the dipper I made for her. They did not come near me again that day. I had now been here six days but I was sure I was making progress. If only I could get the old man to come over to me, get him to eat a full can of snuff; that would kill him for sure, and that way kill himself, I wouldn't be guilty of murder.

"The old lady was a meek old thing. The young fellow was by this time quite friendly. The girl would not hurt anybody. Her chest was flat like a boy—no development like young ladies. I am sure if I could get the old man out of the way, I could easily have brought this girl out with me to civilization. But what good would that have been? I would have to keep her in a cage for public display. I don't think we have any right to force our way of life on any people, and I don't think they would like it. (The noise and racket in a modern city they would not like any more than I do.)

"The young fellow would have been between 11-18 years old, about seven feet tall and might weigh about three hundred pounds.

His chest would be about 50-55 inches, his waist about 36-38 inches. He had wide jaws, narrow forehead that slanted upward, round at the back, about four or five inches higher than the forehead. The hair on their heads was about six inches long. The hair on the rest of their bodies was short and thick in places. The womens' hair was a bit longer on their heads and the hair on their forehead had an upward turn like some women have—they call it bangs—among women's hair-do's nowadays. The old lady could have been anything between 40-70 years old. She was over seven feet tall. She would be about 500-600 pounds.

"She had very wide hips and a goose-like walk. She was not built for beauty or speed. Some of those lovable brassieres and uplifts would have been a great improvement on her looks and her figure. The man's eyeteeth were longer than the rest of the teeth, but not long enough to be called tusks. The old man must have been near eight feet tall. Big barrel chest and big hump on his back—powerful shoulders, his bicep[s] on [his] upper arm were enormous and tapered down to his elbows. His forearms were longer than common people have, but well proportioned. His hands were wide, the palm was long and broad and hollow like a scoop. His fingers were short in proportion to the rest of his hand. His fingernails were like chisels. The only place they had no hair was inside their hands and the soles of their feet and upper parts of their nose and eyelids. I never did see their ears, they were covered with hair hanging over them.

"If the old man were to wear a collar, it would have to be at least thirty inches. I have no idea what size shoes they would need. I was watching the young fellow's foot one day when he was sitting down. The soles of his foot seemed to be padded like a dog's foot, and the big toe was longer than the rest and very strong. In mountain climbing all he needed was footing for his big toe. They were very agile. To sit down they turned their knees out and came straight down. To rise they came straight up without use of their hands or arms. I don't think this valley was their permanent home. I think they move from place to place, as food is available in different localities. they might eat meat, but I never saw them eat meat, or do any cooking.

"I think this was probably a stopover place and the plants with sweet roots on the mountainside might have been in season this time of year. They seem to be most interested in them. The roots have a very sweet and satisfying taste. They always seem to do everything for a reason, wasted no time on anything they did not need. When they were not looking for food the old man and the old lady were resting, but the boy and the girl were always climbing something or some other exercise. His favorite position was to take hold of his feet with his hands and balance on his rump, then bounce forward. The idea seems to be to see how far he could go without his feet or his hands touching the ground. Sometimes he made twenty feet.

"But what do they want with me? They must understand that I cannot stay here indefinitely. I will soon run out of grub, and so far I've seen no deer or other game. I will soon have to make a break for freedom. Not that I was mistreated in any way. One consolation was that the old man was coming closer each day, and was very interested in my snuff. Watching me when I take a pinch of snuff. He seems to think it useless to only put it inside my lips. One morning after I had my breakfast both the old man and the boy came and sat down only ten feet from me. This morning I made coffee. I had saved up all dry branches I found and I had some dry moss and I used all the labels from my cans to start a fire.

"I got my coffee pot boiling and it was strong coffee too, and the aroma from boiling coffee was what brought them over. I was sitting eating hard-tack with plenty of butter on, and sipping coffee. And it sure tasted good. I was smacking my lips pretending it was better than it really was. I set the can down that was about half full. I intended to warm it up later. I pulled out a full box of snuff, took a big chew. Before I had time to close the box the old man reached for it. I was afraid he would waste it, and only had two more boxes. So I held on to the box intending him to take a pinch like I had just done. Instead he grabbed the box and emptied it in his mouth. Swallowed it in one gulp. Then he licked the box inside with his tongue.

"After a few minutes his eyes began to roll over in his head, he was looking straight up. I could see he was sick. Then he grabbed my coffee can that was quite cold by this time, he emptied that in his mouth, grounds and all. That did no good. He stuck his head between his legs and rolled forwards a few times away from me. Then he began to squeal like a stuck pig. I grabbed my rifle. I said to myself, 'This is it, if he comes for me I will shoot him plumb between his eyes.' But he started for the spring, he wanted water. I packed my sleeping bag in the pack sack with the few cans I had left. The young fellow ran over to his mother. Then she began to squeal. I started for the opening in the wall—and I just made it. The old lady was right behind me. I fired one shot at the rock over her head.

"I guess she had never seen a rifle fired before. She turned and ran inside the wall. I injected another shell in the barrel of my rifle and started downhill, looking back over my shoulder every so often to see if they were coming. I was in a canyon, and good traveling and I made fast time. Must have made three miles in some world record time. I came to a turn in the canyon and I had the sun on my left, that meant I was going south and the canyon turned west. I decided to climb the ridge ahead of me. I knew I must have two mountain ridges between me and salt water and by climbing this ridge I would have a good view of this canyon, so I could see if the Sasquatch were coming after me. I had a light pack and was making good time up this hill. I stopped after to look back to where I came from, but nobody followed me. As I came over the ridge I could see Mt. Baker [Mt. Churchill?]. Then I knew I was going in the right direction.

"I was hungry and tired. I opened my pack sack to see what I had to eat. I decided to rest here for a while. I had a good view of the mountainside, and if the old man was coming, I had the advantage because I was above him. To get me he would have to come up a steep hill. And that might not be so easy after stopping a few 30-30 bullets. I had made my mind up this was my last chance, and this would be a fight to the finish. I ate some hard tack and I opened my last can of corned beef. I had no butter, I forgot to pick up my butter sealer I had buried near my camp to keep it cold. I did not dare to make a fire. I rested here for two hours. It was 3:00 p.m. when I started down the mountainside. It was nice going, not too steep, and not too much underbrush.

"When I got near the bottom I shot a big blue grouse. She was sitting on a windfall, looking right at me, only a hundred feet away. I shot her neck right off.

"I made it down to the creek at the bottom of this canyon. I felt I was safe now. I made a fire between two big boulders, roasted the grouse, made some coffee and opened my can of milk. My first good meal for days. I spread out my sleeping bag under a big spruce tree and went to sleep. Next morning when I woke up, I was feeling terrible. My feet were sore from dirty socks. My legs were sore, my stomach was upset from eating that grouse the night before. I was not sure I was going to make it up that mountain. It was a cloudy day, no sun, but after some coffee and some hard tack I felt a little better. I started up the mountainside but had no energy. I only wanted to rest. My legs were shaking. I had to rest every hundred feet. I finally made it to the top, but it took me six hours to get there. It was cloudy, visibility about a mile.

"I knew I had to go down hill. After about two hours I got down to the heavy timber and sat down to rest. I could hear a motor running hard at times, then stop. I listened to this for a while and decided the sound was a gas donkey [engine used to yard logs on hills]. Someone was logging in the neighborhood. I made for this sound, for if only I can get to that donkey, I will be safe. After a while I hear someone holler 'Timber' and a tree go down. Now I knew I was safe. When I came up to the fellows, I guess I was a sorry sight. I hadn't had a shave since I left Toba Inlet, and no good wash for days. When I came up out of the bushes, they kept staring at me. I asked where the place was and how far to the nearest town. The men said, 'You look like a wild man, where did you come from?'

"I told them I was a prospector and was lost. I had not had much to eat the last few weeks I got sick from eating a grouse last night, and I am all in. The bucker called to his partner, 'Pete, come over here a minute,' Pete came over and looked at me and said, 'This man is sick. We had better get him down to the landing, put him on a logging truck and send him down to the beach.' I did not like to tell them I had been kidnapped by a Sasquatch, as if I had told them, they probably have said, 'He is crazy too.' They were very helpful and they talked to the truck driver to give me a ride down to the beach. Pete helped me get up into the truck cab and said, 'The first aid man will fix you up at the camp.' The first aid man brought me to the cook and asked, 'Have you a bowl of soup for this man?' The cook came and looked me over. He asked, 'When did you eat last, and where did you come from?' I told him I had been lost in the wood. 'I ate a grouse last night and it made me sick.'

"After the cook had given me a first class meal, the first aid man took me to the first aid house. I asked, 'Can you get me a clean suit of underwear and a pair of socks? I would like a bath, too.' He said, 'Sure thing, you take a rest and I will fix all that. I'll arrange for you to go down to Sechelt when the timekeeper goes down for mail.' After a session in the bathroom the first aid man gave me a shave and a hair trim, and I was back to my normal self. The Bull of the

Woods [logging boss] told me I was welcome to stay for a day and rest up if I liked I told him I accepted his hospitality as I was not feeling any too good yet. I told about my prospecting, but nothing about being kidnapped by a sasquatch.

"The following day I went down from this camp on the Salmon Arm Branch of Sechelt Inlet. From there I got the Union Boat back to Vancouver. That was my last prospecting trip, and my only experience with what is known as Sasquatches. I know that in 1924 there were four sasquatches living, it might be only two now. The old man and the old lady might be dead by this time." [S24A-map11]

In concluding Ostman's account, Green makes the following observations: "That is the story. If someone told it now it would probably be laughed off, even by sasquatch enthusiasts, because detailed information is readily available in print and several people have made up slightly similar accounts of adventures with the hairy giants. For Albert Ostman, there was no pattern to follow. Some of the things he said of the sasquatch have not been confirmed by the hundreds [now thousands] of later reports. No one else, for instance, has described anything like bark and moss blankets. But his descriptions of the creatures themselves, which were at variance with the common impression at that time, have been confirmed over and over again."[6]

What was the common impression at the time? It seems that in the 1950s and 1960s, sasquatches (and bigfoot) had not lost an aura of "humanness" that Burns and others had first allowed them in the 1920s. For Ostman to come right out and state that they seemed more apelike than human would certainly have run dead against the then-popular image. Other observations that many people still have difficulty with is the notion of "weaving" by sasquatches. In fact, since Green's writing, nests woven of bark and branches have been found and attributed by some to sasquatches. Their construction involved a simple form of weaving and has occasionally been reported to include moss as Ostman attested. Primate research has recently shed new light on this subject. It is entirely possible that since gorilla nests, as described by most authors, involve sequential bending and layering of saplings to make sleeping structures, in fact "simply woven" as described by some researchers, then a somewhat more evolved ape could do as much. Likewise, the chattering observed repeatedly by Ostman at close, even relaxed, quarters need not resemble the harsh alarm vocalizations so often reported in sudden meetings between man and sasquatches. But this is only one of several questionable behaviors mentioned in Ostman's account.

Ostman's report is all quite fabulous, and there can be no proof in any event, but abduction of humans has been mentioned more than once in older literature, quite apart from more recent documented reports. Abduction of women may have occasionally occurred with viable offspring in the past in different locations as attested by B. C. Natives, with effects on sasquatch characteristics that could conceivably explain differences in some sasquatch, or perhaps human, populations. If we never come to know, it may be quite a good thing.

Harry Colp—The Strangest Story Ever Told

(This account is also known as "The Apemen of Thomas Bay.") If there is one place still feared and shunned in southeast Alaska, for whatever reason, it is Thomas Bay north of Petersburg. In *The Strangest Story Ever Told*,[7] as recounted by the late Harry Colp of Petersburg, Alaska, and later made ready for publication by his daughter Virginia Colp, there are numerous references to unusual occurrences in the mountains behind Thomas Bay. Among them is a report of hair-covered upright creatures. And in the years between 1900 and 1919, there were a number of mysterious and disturbing events experienced by the men who went back looking for gold.

Thomas Bay is located on the mainland, halfway between Petersburg and Juneau to the north, in the rugged, glacier-studded mountains of the southeast Alaskan panhandle. Shrouded in hemlock and cedar, it was long the location of rumored gold finds, but generally avoided by Tlingits. Around 150 years ago, it was the scene of a terrible tragedy, when a slide from one of the mountain slopes obliterated a Tlingit village, killing more than 500 inhabitants. Since then it has been known in Tlingit as "The Bay of Death."

In his book, Colp recounts how he and three others from Wrangell ventured there by boat in 1900 for the purpose of finding a rumored gold lode, about eight miles up the Patterson River on the east side of Thomas Bay. The other men were referred to by Colp as John, Fred and Charlie. The book contains accounts of the initial trip in the spring of 1900, another by John and Fred in July, 1900, and subsequent tips by various members of the group in 1903, 1906,

1908, 1911, 1914 and, finally, 1919. Following are some excerpts from the book with possible connection to sasquatches.

Spring, 1900, in the mountains southeast of Thomas Bay, recounted by Charlie:

"Swarming up the ridge toward me from the lake were the most hideous creatures. I couldn't call them anything but devils, as they were neither men nor monkeys—yet looked like both...their bodies covered with long, coarse hair, except where the scabs and running sores had replaced it. Each one seemed to be reaching out for me and striving to be the first one to get me. The air was full of their cries and the stench from their sores and bodies made me [feel?] faint.

"I forgot my broken gun and tried to use it on the first ones, then I threw it at them and turned and ran. God, how I did run! I could feel their hot breath on my back. The smell...was making me sick; while the noises they made, yelling, screaming and breathing, drove me mad..."

Charlie then described how he managed to escape, getting down to his canoe and paddling along the shore back to civilization, where he told his story to Colp and then left Alaska for good. The men passed off Charlie's story and in July, 1900, John and Fred returned to Thomas Bay. John recounted to Colp how they had made their way up the Patterson River searching for the landmark lakes that would lead them to the gold. Before parting they camped on a big tundra flat at 2,500 feet elevation, dotted with scrubby jack pines, and then separated. After a short trip apart, John was returning down the mountainside when he heard three rapid shots.

Returning, John found Fred quite incoherent, noting that "the place was dug up in small holes for quite a distance, several small pine trees were pulled up and scattered around, and to one side of the dug-up area was Fred's gun stuck down in the tundra up to the hammer." He pulled it out and found one empty shell in the barrel and three still loaded in the magazine. John finally persuaded the dazed Fred to leave with him and made a camp down below. After supper in the tent, Fred muttered something about the creatures being after him, grabbed his gun and took off for their two boats. Colp added: "John listened and thought he heard the sound of padded feet," and without collecting all of their prospecting gear, John followed quickly behind Fred, eventually making their way back to Wrangell. Colp later persuaded John to go back with him for the tools, blankets, cooking gear and powder left there; they made a short and uneventful trip, collecting their gear and returning safely.

Both men returned again in 1903, and prospected around for a few days finding iron deposits, until, in Colp's words, "John was anxious to get out for some reason, and for myself, I didn't feel any too comfortable. It seemed as if someone or something with a menacing disposition was right behind me all the time, and that there were hundreds of angry eyes watching me. I had a feeling of danger, but kept my mind forced on the affairs at hand."

The two men agreed there was a strange presence to the place and left, spending an uneasy night back at base camp. Colp continues: "That night John awakened me by nudging me in the side and whispering 'Harry, wake up. Do you hear that? Wake up! Listen, can't you hear anything?'

"Coming out of a sound sleep knowing what I did about that country and having a fellow keep whispering are enough to make anybody's nerves tingle. Slowly, through my sleep drugged senses, came a queer sound—half squeak and half moan—which seemed to be within a few feet of my hearing. The noise of chips or small branches being broken could be heard. I sure was scared. Every hair on my body stood on end, and cold shivers ran down my spine. I sat up in bed, grabbed my gun, and threw a shell into the chamber while I listened."

Colp concluded that it was just some unseen animal that had unnerved John, but noted: "There was no sleep for him that night, and just as soon as morning came, we broke camp and started back for Wrangell." Colp added that although they would often get out together again after that, John never would go back to Thomas Bay.

Colp went in again in 1906, with another gold seeker. On the way the new partner had a fall and had to be helped out, cryptically refusing to ever go back. In 1908 Colp ventured back alone, with supplies for quite a stay. He anchored up at a nice little cove and noticed a deserted tent and boat up the beach. There he found a rusting rifle, supplies of all sorts and evidence that no one had been there for six to eight weeks. Finding this deserted camp was unnerving to him, and he left, determined never to come back alone. Colp reported the missing man, no trace of whom was never found.

In 1911 Colp made his last trip in with a big Norwegian named Ole, but they returned with no gold and a good deal of unease on

Ole's part. In 1914 Colp tried sending two other men in, but they wasted no time in returning, saying they had had enough of Thomas Bay. Colp again sent men in 1919, and they all returned with conflicting stories. The final story in his book ends with a somber and mysterious account of yet another missing man.

A dairy farmer Colp knew and believed, told him that at his homestead, about "sixteen miles from Devil's country," an old trapper had camped about a mile upriver from him during the winter of 1925. The farmer told Colp the man had run a trap line up Muddy River as far as he could travel in one day, and several side trap lines out from there, one near the landmark lake. The man had found some quartz showings there that winter and had camped out. Colp relates the trapper's ominous story in the farmer's words, heavy with foreboding.

"That night it snowed about six inches. During the night he heard his dog bark, and he thought he heard it howl. Once it seemed to run away, barking as it did. In the morning, no dog was in sight. He thought it seemed funny, because the dog had always been around camp in the mornings, no matter how much he roamed at night. He couldn't have gotten caught in a trap, because he was trap wise.

"He took the dog's trail and followed it a little ways from the tent, when he came to the same strange tracks following the dog's. The dog's and the strange tracks were all mixed up for about five hundred yards, when the dog's tracks disappeared and only the strange tracks went on.

"He stood and studied the tracks but could make no sense of them. The dog tracks were going along and there were the strange tracks going right after them. In some places the mysterious owner of the tracks had stepped into the dog tracks and covered them, showing that it was behind the dog. The dog tracks showed that the dog had been running, but when they ended there were no signs of a struggle. It looked as if the dog had vanished or had been picked up clear of the ground right in the midst of its steps.

"The trapper, after looking the ground over, decided he would follow the strange animal's tracks and see where it had gone. If he could find its den or where it had stayed, he might get a chance to shoot it and see what kind of animal it was. So off he went.

"The tracks went in a big circle and after a time came back to where the trapper had taken them up at the original place where the dog tracks left off and now were following the owner. Twice they went around in the same circle, but no sight of the creature could he get.

"He began to think that this was no place for him, so he went out to the farmer and asked what kind of animal was in that section of the country. Its tracks were of two sets: the hind set were about seven inches long and looked as if they were a cross between a two-year-old bear's and a small bare-footed man's tracks.

"You could see claw marks at the ends of the toes, toe pads and heavy heel marks; between toe-pad marks and heel marks was a short space where the foot did not bear so heavily on the ground, as if the foot were slightly hollowed or had an instep. The front set looked like a big raccoon's tracks, only larger [a raccoon forepaw has five digits, the "thumb" is partly opposable]. Sometimes the hind set were all that could be seen, as if the creature could and did walk sometimes on all fours and sometimes on only its hind legs.

"The farmer told him he knew of nothing like that ever being seen or heard of around there. He himself had seen nothing like that in ten years he had been around.

"The trapper went back and disappeared. Three weeks after that, his outfit was found. Traps on the trap line were in various positions—some set and others sprung. No trace of the trapper was ever found." [OS00-map2]

So ends the final chapter in Harry Colp's book, a manuscript which he had prepared but never published before his death. Through the efforts of his family, the manuscript has been published numerous times, the most recent of which is noted below. It is an intriguing short book and is worth reading in its entirety.

What are referred to as "neither monkeys nor men—yet looked like both," may or may not have been a reference to hair-covering and facial anatomy of anthropoids in general and the erect posture seen in *Homo sapiens*. Since at no time in the story is there any anatomical reference to tails, it may be that "monkey" might have been meant in the expanded context of apes and monkeys, as was occasionally the norm in past times. It should be noted that the term "devil" is often found in early American newspaper accounts of sasquatchlike creatures and, as such, may simply be a reference to sasquatches in the vernacular of the times. How big Charlie's hairy creatures were, one cannot tell. The exact size and details of the tracks following the trapper's dog remain equally puzzling. The

Colp story, quite simply, defies any easy explanation, but remains today one the best-known old stories in southeast Alaska, often referred to as "the Apemen of Thomas Bay."

Watson's "Little Hairy Devils" of Hecata Island

Stories of manlike creatures considerably smaller than a sasquatch have in the past hundred years been noted less frequently than those of larger creatures; but, even in the last century, they have refused to fade away entirely. In assembling a collection of sasquatch narratives from Prince of Wales Island, Al Jackson of Klawock was given the following story by his uncle, the late Mr. Al Anniskette, also of Klawock. It concerned a certain seasonal fish packer-boat operator and prospector from the "lower forty-eight" named Cowboy Watson and a number of small, quite aggressive, hairy creatures.

"My uncle told me that, in the late 'twenties, Watson would work for him regularly each year during salmon season, packing fish on his boat for the Klawock cannery, and save up enough to outfit himself for a full fall season of gold prospecting on west Hecata Island. Hecata is located off the northeast corner of Prince of Wales Island, about sixty miles north of Klawock. Watson would come back each year towards Christmas, whenever the snow would put an end to his gold recovery. He would do well each year and was able to bank all of his gold. He did well at it and showed a good profit each time he would go south.

"In the fall of '28 he came out early, looking 'real beat up.' His clothes were all in tatters and he didn't say much. He just went to the office straightaway and got his ticket south. He had his gold, but he said to me, 'That's it, somebody else can have it, I'm not going in there any more.' When pressed for details he simply said, 'Those little, black, hairy devils drove me out of there. They're small, maybe not much bigger than chimpanzees, three and a half, maybe four feet high—but man, are they tough!' Watson looked like he'd been in a real fight. He added again, 'That's it, I've had it. I won't be coming back,' and headed south." [OS48-map3]

Mr. Jackson was personally familiar with the Thomas Bay account of Harry Colp's, and added, "I knew my uncle didn't read much, if at all, so I prodded him, saying,'You know, that sounds like something I've heard from Thomas Bay.' But he just looked surprised and said to me, 'I've never heard anything about any Thomas Bay.'"

This last story attributed to Cowboy Watson on Hecata Island in 1928 is in many ways similar to Colp's account of the early 1900s, especially in the smaller size of creatures suggested, as well as the extreme aggression alleged. Unless the creatures described in both accounts were juvenile sasquatches, the nature of the reports suggest a hominoid of a different type.

Tlingits have traditionally attested the existence of a man-sized creature, the *kushtakaa*, or land-otter man, a "human gone wild" and supernaturally transformed, much feared fifty or sixty years ago. Some Native authors have suggested that historically, juvenile or subadult sasquatches may have been responsible for some older accounts that seem to fix on a man-sized hairy hominid. But there have also existed, among traditional northern Tlingits, stories of much smaller creatures such as the "Dwarves of Pybus Bay." Pybus Bay is located on the southeast corner of Admiralty Island, home of the largest brown bears in southeast Alaska. The late author and coastal navigator, Stephen E. Hilson, cryptically noted the following about traditional beliefs surrounding the area.

"[Tlingit] Indians have recognized Pybus Bay as the home of little dwarfs who although they chose not to reveal themselves, were very helpful to men. They were small, but so heavy that no one could lift them."[8] Such smaller legendary creatures are no better at matching the little aggressive types than regular sasquatches, and leaves the whole "little" or "aggressive" affair in a bit of a muddle.

In all of these preceding accounts there may be something more than just the spinning of yarns. The story of Frank Howard, while certainly quite startling by the standards of his time, is really not that different from any late twentieth-century sasquatch report, excepting perhaps the light bluish gray color of the creature reported. Only one other southeast report mentions a gray sasquatch. Then again, Yakutat is further north, and according to Bergman's Rule, a sound tenet of modern biology, mammals in northern climates are more likely to be of a lighter color than those of warmer climates. John Green has also mentioned a gray sasquatch reported inland from Yakutat, just northeast of the panhandle. That Howard's Yakutat creature was alleged to be curious but not threatening is also well documented behavior in many current reports from Alaska and elsewhere.

Ostman's long story is still remarkable, years after interviews by Green, Sanderson and Dahinden, who were all impressed with the man's consistency and forthright manner. Green detailed the few apparent inconsistencies in a most impartial and quizzically skeptical fashion and his comments on the story should be read first hand. Green especially points out that many of the sasquatch attributes Ostman detailed, while unheard of when the story was first told, are now repeated individually in recent reports. The blankets of bark strips and moss mentioned are still unmentioned in current reports of sasquatches, but it may be a small jump from what some zoologists have called the simple "woven" appearance of gorilla nests to the interlacing of material reported in some possible sasquatch nests. Otherwise, Ostman's sasquatches are really an excellent synthesis of what are now regularly reported sasquatch attributes unheard of in previous literature, even including a chattering form of close communication. Other characteristics, such as carrying foodstuffs and a primarily vegetarian diet, accord well with modern reports. As to Ostman's "being picked up," there are about a half dozen reports of people being picked up and carried off. One such report noted by Burns and Green and summarized by Bord,[9] describes a B.C. Salish Indian woman named "Grandma Charlie," who, in 1871, "claimed she had been kidnapped by a Bigfoot and forced to live with it for a year. She was seventeen at the time and living in the Fraser Valley. The male creature is said to have forced her to swim the Harrison River and then carried her to a rock shelter where it lived with its parents. It kept her for a year, but then took her home because she 'aggravated it so much.'" In all reports she attested that she was not treated unkindly—astonishing perhaps, but hardly unique.

The fact is that accounts of abduction by sasquatches are not new by any means, far predating even the tabloids. John Green and veteran researcher Peter Byrne both interviewed a priest on the western coast of Vancouver Island who knew, first-hand, the 1928 story of "Muchalat Harry," a Native trapper who was allegedly carried off by a tall sasquatch to a place where a group of the creatures was gathered. He was said to have later escaped by traveling overland twelve miles to his canoe, then paddling back to his village. Colin and Janet Bord describe at least four other instances of alleged abduction or attempted abduction of humans in North America by sasquatchlike creatures.[10] The very event of kidnapping, although mentioned in at least two other detailed accounts, may suggest something more. Following one interview with Ivan Sanderson, Ostman somewhat shyly confided the suspicion that he had been kidnapped as a potential suitor for the young female.

This sort of notion harkens back to numerous stories from Natives and older tribal beliefs as well, involving two second-hand accounts from Nanaimo, which I obtained from older Nanaimo Native informants, of abductions of women being followed by the birth of half-human, half-sasquatch infants. In one story, the infant was said to have been stillborn; in the other, the infant died shortly after birth. In both accounts the bodies were said to have been slipped off a ferry at night, midway between Nanaimo and Vancouver. One event was supposed to have happened to the grandmother of an elderly woman living in Nanaimo in 1974, both events said to have occurred around the early 1900s.

The most unsettling of all these stories is the Thomas Bay story of Harry Colp, yarn or not. The creatures described therein certainly appear to resemble small sasquatches, and the many unexplained events are suggestive of a mainland Alaskan population of some species that could not be easily called a "wild human." While the actual of description of the creatures as "not men" and "not monkeys" but something "in between" is scanty. The attributes of hairy bodies and aggressive behavior when confronted with the invasive presence of humans in an unpopulated and rugged spot are each documented in numerous sasquatch reports from the Pacific Northwest. The mysterious disappearance of the trapper's dog matches curiously with the well-documented enmity between sasquatches and dogs. And the trapper's own equally unsettling disappearance soon after, if true, matches two other such accounts from northern Alaska where "bush-men" were believed by Athabascan Natives to have killed men. One does not know to what extent, if true, such alleged attacks might have been precipitated by the humans involved—shooting at such creatures, for example.

Reports of what appear to be sasquatches from all these earlier coastal accounts basically serve to illustrate a range of characteristics that are also common to man: a variation in size and color. At the same time it may be noted that most all of the behaviors, the rare alleged abduction of humans excepted, are those same behaviors found among the great apes. In the chapters that follow, attention is given to tracks and other purported evidence. While attributes such as erect posture, a somewhat "flat" face, good swimming ability and occasional reports of complex vocal patterns all suggest a possible relationship to our species, the suggested anatomy of the foot, some of the more "apelike" vocalizations and casual behavior toward man all suggest a hominoid that is not typically human in most ways. In fact, it would appear in many ways to be just as closely related to the great apes.

Then again, it may be that sasquatches are neither "apes" nor "human," but simply a third family on their own— refusing to conform to scientific ideas of what constitutes a pongid or a hominid— and instead following their own path, one that leads inexorably into the nighttime forest.

1. Mack, Clayton, 1993, *Grizzlies and White Guys*, Harbour Publishing, Madeira Park, B.C.

2. Mack, 1993, pp. 123–125.

3. Mack, 1993, pp. 128–130.

4. Howard, Frank, 1908, *The Alaska-Yukon Magazine*, March edition, Vol. V., no. 1.

5. Ferrell, Ed, 1996, *Strange Stories of Alaska and the Yukon*, Epicenter Press, Kenmore, WA, pp. 16–18.

6. Green, John, 1980, *On the Track of the Sasquatch,* Hancock House Publishers, pp. 11–18 (originally published 1968).

7. Colp, Harry, *The Strangest Story Ever Told,* Exposition Press, New York, NY, ©Virginia Colp.

8. Hilson, Stephen E., 1997, *Exploring Alaska and British Columbia, Skagway to Barkley Sound*, Evergreen Press, Seattle, WA, p. 68.

9. Bord, Janet and Colin Bord, 1982, *The Bigfoot Casebook*, Stackpole Books, Harrisburg, PA, p. 24.

10. Bord and Bord, pp. 60, 125, 159, 176.

Illustration: Patrick Beaton

Kwakiutl pole of D'sonoqua, Wild Woman of the Woods, topped by supernatural eagle.
Photo: David Hancock

D'sonoqua mask. Photo: David Hancock

Right page: D'sonoqua mask and suit danced by Tsungani at Tlingit Chief Don Assu's potlatch. Carved by Don Lelooska.
Photo: David Hancock

D'sonoqua doll—*Wild Woman of the Woods*—carved by Lelooska. She is carrying off a basket of children. *Photo: David Hancock*

Right page: Kwakiutl D'sonoqua pole from Totem Pole Park, Victoria. *Photo: David Hancock*

Above and left page: *Illustrations by Patrick Beaton*

Illustrations: Patrick Beaton

10 Native Sasquatch Folklore

Traditional Native perspectives of the sasquatch and creatures like them have a useful purpose for any wildlife biologist, zoologist or anthropologist investigating sasquatches in the coastal Pacific Northwest. In interviewing witnesses in any area, it is often useful to know "where people are coming from" in their assumptions about any such creatures. The following perspectives may be viewed either as themes concerning various sasquatchlike creatures or even as possible physical and behavioral attributes of real species of undiscovered hominids, living or extinct.

Tlingit Beliefs

The Tlingit occupy most of what is today termed southeast Alaska. There, the belief in hair-covered manlike beings called *kushtakaas* is still strong today. Although similar to a man-sized sasquatch, they were (are) believed to be the embodiment of lost relatives. As related to me by numerous Tlingit informants from 1992 to 1999, *kushtakaas* were essentially men who through near drowning or becoming lost had been influenced by the *kushta* or land otters and had grown hair all over their bodies, lost their reason and gone to live with others so transformed—the "land otter people."

Traditionally, *kushtakaas* were able to live off the sea life abundant along the rocky shores and beaches and, like the otter, whistled to communicate with each other. They were also attributed with a call of "something like a baby crying." They were intensely feared for their ability to capture one's soul by hypnosis.

In early texts on Tlingit folklore, wars with the *kushtakaas* were frequently mentioned, with either side the victor depending on the account. Tlingit doctors, *ixt,* derived their power through the otter,

Figure 21: Photograph of a ceremonial hat representing Tlingit *kushtakaa*, or "land-otter man."

especially its tongue, and such individuals were the foremost defense or cure for a captured relative.

Killing a *kushtakaa* was expressly forbidden as it may likely have been a lost relative, or since killing one might have incited a war with the *kushtakaas*. Of all Native people in southeast Alaska, the Tlingit especially have continued to carry a strong regard for leaving such creatures alone, and would certainly take issue with anyone inclined toward shooting such a being, although such Native events are occasionally referred to among the northern Tlingit.

Kushtakaas were variously portrayed in carvings as beings in human form, but hairy and sometimes down on all fours, as if in transformation into a land-otter. In most older accounts, the arms were said to come out of the chest—perhaps an allusion to an extremely round-shouldered posture. A short or long tail was sometimes mentioned, but only occasionally.

The carvings of *kushtakaas* do not reveal a tail. *Kushtakaas* are still strongly believed in by many, if not most, tribal elders, and the sound of a baby crying in the forest at night or peculiar whistling has to my knowledge caused at least one respectable Saxman, Revilla' Island woman to seek the spiritual support of her church. It is traditionally accepted that such omens may portend sickness or worse in one's household or among friends.

When speaking Tlingit aloud it was not customary, except when going to war or expecting one's imminent death, to refer to *kushtakaas* with the euphemism, "We who live at the point" (meaning *kushtakaas*) "welcome death as food for the soul." This phrase was extremely revered and poignant and was actually one's own eulogy, taking on a strongly spiritual significance and not to be taken lightly. However one may wish to look at *kushtakaas* in Tlingit society today, whether as real animals similar to the sasquatch, or humanlike beings with spiritual connections with man, they have a powerful reality on many levels which is still wise to respect.

Ethnographer John R. Swanton recorded numerous Tlingit myths and legends, which he published in 1909. One such tale was the following that relates to *kushtakaas*, excerpted from *Tlingit Myths and Texts* recorded in Sitka under the title "Various Adventures in Cross Sound." It involved several hunters who had broken the taboo of killing a land otter.

"After the pot had boiled and they began eating [a land otter they had killed], something began to whistle in a tree near by and threw a rock down. They threw one back and soon rocks were flying back and forth. It was [not] a great thing to fool with. By and by the men said, 'You might cut our faces,' so, instead of throwing rocks they seized long cones and threw these back and forth all night. Toward morning the being in the tree, which was a land-otter man, began to hit people, and they on their part had become very tired. Finally they tried to get him down by lighting a fire under the tree where he was sitting. When it was burning well, all suddenly shouted, and he fell into it. Then they threw the fire over him and he burned up. But when they started for the beach to go home, all wriggled from side to side and acted as if they were crazy; and when anyone went to that place afterward he would act in the same manner."[1]

A good example of traditional Tlingit belief in the transformational aspect of *kushtakaa*, was given me in 1999 by a Petersburg Native, Herbert H. His story was based on an experience that his uncle had related to him, occurring on a gravel road on the outskirts of Petersburg in the summer of 1948, or about that year.

"Late one afternoon my uncle was walking down the old dirt road south of town when he heard a creature moving in the roadside bushes to his right. He would catch glimpses of a hairy, man-sized creature on his right, then another to his left. When he would speed up, they would speed up. When he would slow down, so would they. It gave him the creeps. So he really picked it up and he could see the one on his left moving along beside him, occasionally looking at him. What nearly popped him out of his skin was that the one on his left would mimic his facial expressions and then much to his surprise, the one on his left appeared to transform its face into his very own facial likeness!" Herbert added that his uncle said "'It really spooked me!'"

Herbert concluded, "You probably know Tlingit folklore acknowledges that *kushtakaas* can change shape into people, that's how they try to lure you into the forest. They'll try to hypnotize you with their thoughts, and they'll try to get you to become one of them."

Klawock Thlingit researcher, Al Jackson, [he prefers the "Th" spelling] who has also lived in Saxman, Revilla Island, has pointed out that among southern Tlingit: "Besides *kushtakaas* or 'land-otter men,' which were often-regarded as spiritual beings, in Thlingit there is another name for bigfoot—*hootsland,* literally meaning 'bear-man.'" There are still some people in Ketchikan and Saxman today who recognize this term as describing a hairy, man-shaped creature in the bush, not to be confused with the older beliefs in *kushtakaas* and perhaps more resembling a generic sasquatch.

Haida Beliefs

The Haida traditionally lived on Haida Gwaii (also known as Graham Island in B.C.'s Queen Charlotte Islands) as well as on the southern half of Alaska's Prince of Wales Island. The Alaskan Haida are now principally congregated in the village of Hydaburg on the island's southwest coast and at the small village of Kasaan on the central east coast. Traditional Haida stories from southeast Alaska also reflected the belief that a man lost in the bush may be transformed into a wild person. This is virtually the same belief as among the Tlingit. The respected Haida elder and tribal historian Mr. Robert Cogo related this story as directly quoted from a Ketchikan Native student yearbook *YAK'EI* in 1984.

"*Gagiit* in Haida, is something similar to the Tlingit *Kushtakaa*. Evening story telling time was not complete without a *Gagiit* story. It put a little fear and caution into the Haida youth. The season for a

Gagiit doing usually involved after a person was lost at sea or only his capsized canoe was found. When unable to find a body the possibility of the lost man turning *Gagiit* gained wide acceptance. Soon, someone out hunting or camping would hear unusual sounds would find a large foot- and forearm-prints in the mud or ground. A *Gagiit* was at large. Relatives of the lost one were alerted. We could call such a man *Gagiit* or "man on all fours" (*Hlgadahldaa*) which in Haida means a man falling from fatigue, cold and hunger to become a wild man. Human to *Gagiit* was complete. I have this story from my late father Eddie Cogo (*Dushwan*) who died in 1946 at the age of 86. He was a great Haida storyteller (*K'iigang*) and (*Gyhlang*). He was a Raven (*Yaahl*) and was of the Brown Bear clan (*Xuuts*) and (*Xuutsnay*) Brown Bear House.

"A *Gagiit* was very elusive and seen from a distance only rarely. Word came that one was seen on Mabel Island. A plan was agreed on. Six clan members were to be secluded and hidden on the island beach. Their faces were to be blackened with ashes (*gayt*).

"The first night was very cold. This was the month of October (*Canuut*). The second night was the same. Morning came and soon the sun was coming up nice and warm above Mable Bay. A large noise was heard which was very weird. Everyone hidden on the edge of the beach (*Chaaw sallii*) became tense. Suddenly an odd looking figure came into view out of the woods. The figure was hard to look at. What once was a human being (*Nang xaataa*) was now a weird-looking wild man. A short time later the figure continued on down the beach and seated itself on a large flat rock. It was seeking warmth from the morning sun. It was soon decided to put the *Gagiit* out of its misery (*laanadaaang*). The sharpshooter for the group raised his musket (*juk*) and fired, killing the *Gagiit*. The creature was buried right on the edge of the beach and to this day this place is tabu to Haidas. Those trying to trap or hunt there get strange feelings so they have kept away. The spirit of *Gagiit* is there."

After recounting this early 1900s story from his father, Mr. Cogo continued.

"When I was fifteen years of age, I saw something very similar. A group of hand trollers (*xaawla' aayggaa*) were taking it easy before a large campfire on the beach. Someone yelled 'Look!' A strange long-haired face was looking out of the trees and brush across the small bay. Immediately about fifty men circled the wood to try and catch whatever it was. It was a man who had been reported missing for over a week. They caught this man. He was in the process of turning *gagiit*. He died almost within the time they caught him. They tried to warm him before the fire, the change was too great. He died. This is the end of my story."

Some accounts of *gagiits* as "wild people" from Haida sources leave one wondering how difficult it must be trying to harvest natural resources with no technology, that is for a human, however inured to hardship by the deprivation of wilderness living. One elderly Haida man, whom I knew for some years before he passed on, Dexter Wallace of Kasaan, told me in 1994 that he had heard a number of things about *gagiit* from other yet older Native commercial fishermen over the years. "For example," he said, "they eat all sorts of things. They don't just eat berries, deer and roots. They can swim real well. When you're on a new beach somewhere, and you find a pile of fresh abalone shells all stacked up, that's where a *gagiit* has gone in for 'em. They've got hair all over and they're real strong. There could still be some around."

Mr. Ray Sanderson of Hydaburg, a practical man of Haida ancestry who grew up being quite familiar with stories of *gagiit*, related something in August, 1999, regarding vocalizations made by the creatures, "I don't think there's a hunter in Hydaburg, who if he were in the bush and heard the sound of a baby crying, wouldn't turn around and head right back the other way!" Vocalizations such as this crying sound, along with chuckling, whinnying and screams or yells have all been attributed to sasquatchlike creatures. Mr. Sanderson also related some brief aspects of traditional beliefs relating to "land-otter man," as he referred to *gagiit*.

"The old people used to say that if you caught one you could tell if it was a *gagiit* by burying it up to the neck in the sand and lighting a fire around it. If mice ran out of its mouth or nose it was a *gagiit*. There were also stories about land-otter women. They were also covered with hair but had webs between the toes."

McNair, Joseph and Grenville, in their excellently illustrated text of carvings *Down From a Shimmering Sky*, summarize the Haida *gagiit* beliefs succinctly. "Among the Haida, those who barely escape drowning are similarly attracted to an elusive light. If they reach the fire they lose all reason and become a *gagiit*. Near starvation, they end up living in the intertidal zone, searching for what

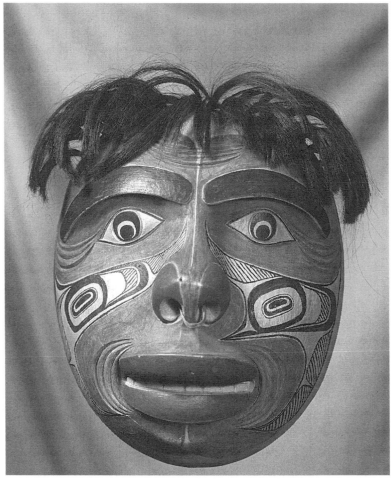

Figure 22: Photograph of a carved Haida mask from Haida Gwaii, B.C., representing *gagiit*. Artist is Charles Edenshaw, 1898. *Photo: American Museum of Natural History*

food they can catch by hand, usually small spiny fish and sea urchins. Because they do not have utensils, they consume their minimal diet raw, and the bones and spines of their prey become permanently embedded in their lips."[2] Some masks do not portray this trait, while others do.

One Haida informant in Hydaburg told me in 1994 that she had seen two *gagiit* while on a boat with her father and crew not far south of Hydaburg close to shore the previous year. She said, "They were just the size of people but covered all over with hair. We all saw them just sitting on the rocks. They were covered with brown hair. My father called them 'Hairy-man.'" She was interrupted by her father entering the room, who indicated his displeasure at her narrative and that he did not wish her to be more forthcoming, for whatever reason. There persists among a good many traditional types, a belief that to discuss *gagiit* or its Tlingit counterpart, *kushtakaa*, will bring bad luck or even death.

Not all Natives hold rigidly to the archetype beliefs. Many Natives today have come to acknowledge the beliefs as part of a truly rich cultural heritage that imbues a profound respect for nature and ethical means of interacting with it. As well, Native stories about all aspects of nature often had embodied in them useful rules for managing a resource or for surviving in a challenging environment.

In recent years, some Natives have given consideration to the idea that legend and folklore regarding "wild man" may have some relation to the thousands of sasquatch reports from western North America. A good example of this approach is the thoughtful perspective offered by the late Haida, Albert J. Brown. In a Ketchikan newpaper, *New Alaskan*, dated October 1988, Mr. Brown[3] devoted a column to "Alaska's Big Foot" and looks critically at the traditional stories and what reality might be found behind them. Brown recounted several interesting gagiit stories but noted, "I believe that the *Gah-Geet* was a species related to the *Bigfoot* of the Rockies. He coincides with the description of *Bigfoot* except for his height. The Rockies' *Bigfoot* is always described as being seven feet tall. But, the local wild man was of average human height which made the legends more believable.

"I think when the musket became an instrument of common use, it spelled doom for the *Gah-Geet*. They apparently did not reproduce in large numbers, so the destruction of a few would lead to the demise of the specie."

Albert Brown was born in 1911 in Alaska, was college educated, and had a successful career as a shipwright supervisor in southeast Alaska; in short, he was a diligent and creditable man. While today Alaskan reports of sasquatchlike creatures still often get no further than immediate family or hunting partner, it is refreshing to see the candor and spirit of scientific inquiry so alive in his narrative. Although Albert Brown was one of the first, he was not the only Alaskan or British Columbian of Native ancestry to step forward with the suggestion that behind the stories there may be a tangible reality, as we shall see later.

Tsimshean Beliefs

Spanning the large country between the Tlingit of southeast Alaska and the Bella Bella of the central B.C. coast is the ethnic nation of the Coast Tsimshean and the related Nisga'a peoples. This includes the areas around the Portland Canal, present-day Prince Rupert, Terrace, and Kitimat, B.C. Since 1900, the Tsimshean have also lived on Annette Island, south of Ketchikan, Alaska. The sasquatchlike *buk'wus* is known in the Tsimshean language as variations of *ba'oosh* or *ba'wes*. This translates, according to the most thorough and recent dictionary of the Tsimshean language compiled by John A. Dunn, to "ape, monkey, Bigfoot, anything that imitates man."[4] While there are other names for "wild man" such as *gyaedem gilhaoli* ("men-of the-woods") and *gyaedem ryaldo,* as noted by Barbeau,[5] *ba'oosh* remains the most universal and is used in both Alaska and B.C. While some Tsimshean may be just as disinclined to speak of *ba'oosh* as Haidas are of *gagiit* or Tlingits of *kushtekaa,* there is less emphasis in their stories, at least today, of a lost man transforming into a wild man. Rather there is somewhat more inclination to regard the creatures in current folk narratives as just another sort of hominid, albeit hairy and without benefit of much technology besides rocks and sticks. In the traditional Tsimshean beliefs, he is a member of a forest-dwelling remnant population of hairy "people" who roam about at the simplest level of existence, but are extremely adapted to a solitary life without the social benefits of human shelter, clothing or tools.

In the following short account from the early 1900s, an elderly Tsimshean, Mrs. N. W., gave her recollection of an interaction between her family and a ba'oosh while they were staying in a cabin on the north end of Annette Island, approximately twelve miles southeast of Ketchikan.

"Around 1935, I was a young girl accompanying my parents on a trip to an old cabin near the remains of Chinatown, a small deserted village site on Annette Point, Annette Island. One night, we started hearing noises like rocks being thrown on the roof of our cabin and my mother got upset, asking my father 'What on earth is that?' At that my father replied calmly, 'Don't worry dear, that's just some old monkey throwing coconuts!' In the morning there was a trail of large tracks of something on two feet leading away from the cabin into the bush."

Chinatown, long deserted, is regarded even today as a place where nobody wants to camp and for years such stories have kept residents of Metlakatla away.

Tsimshean *ba'ooshes* are described often much the same as a sasquatch but are not always seven feet tall. One report from a Native Metlakatla resident described what may have been an infant sasquatch (see [S80B], by Ms. T. D.). Usually information about what to do if one runs into a creature of this sort is passed along in Native families, but generally not publicized outside of the family group for fear of ridicule, Native or non-Native identity notwithstanding.

On Annette Island in 1993, a strapping young commercial fisherman and hunter, D. G., quietly told me of an incident that had happened to him in 1987 while he was dropped off to hunt on the remote southern shore of Duke Island. Duke is a fairly large and uninhabited island to the south of Annette Island, toward Canada. He had seen no sign of other hunters on the way around and did not expect in fact to see a single soul.

"I had made a nice camp at the top of the beach where there was no wind and had the forest to my back. I had my tent set up and had gotten back from scouting out the area. It was getting dark and I had my fire going, it was a calm evening and I was just watching the waves come in.

"All of a sudden I felt something between my shoulder blades and I saw a pine cone roll down beside me. I looked back at the trees and there was nothing to see. Forty feet is too far for a squirrel to throw, if you know what I mean, and I recalled what my dad had told me about *ba'oosh*. 'If you run into one, the best thing to do is just ignore him and he'll go away. They'll play tricks on you like that but one should just ignore them, if you don't bother them, they won't bother you.'

Figure 23: Author's illustration representing Tsimshean or Nishga *ba'oosh* or *ba'wes* mask collected by Emmons in the early 1900s on the B.C. coast. Art based on photographs by Hillel Burger of mask in Harvard Peabody Museum.

"So I just continued sitting there and about five minutes later—*whap*—out came another pine cone. Well, it was like my dad said, I didn't bother with him and that was all that happened, he just went away. I never did see any sign of him, or smell or noise or anything, that was all."

Nuxalk Beliefs

The Nuxalk-speaking Bella Coola live inland and south of the Tsimshean, at Bella Coola and along the fjords just east of Bella Bella, as far south as Namu, B.C. (The Bella Coola word for sasquatch is very similar to the Tsimshean *ba'oosh*. Among the Bella Coola, the name has been recorded by linguists and anthropologists as *bukw's, boq's* occasionally *pukw's*.) Current usage of the term *bukw's* (wild man of the woods) continues today. Masks depicting *bukw's* are fairly consistent up and down the coast but do show some of the differences in styling that each tribal group use to mark the ethnicity of their carvings. Some masks show a hook-nose with wide nostrils, others show *bukw's* with the flattened apelike nostrils of the giantess *dsonoqua* (also variously spelt *tsonoqua, dsonaq'wa, dzonokwa*, etc.).

Of narratives concerning the *boq's*, ethnographer T. W. McIlwraith[6] has the following Native accounts (recorded between 1922 and 1924) to relate.

"Not many years ago a certain *Qak'lis* [man's name] was encamped with his wife and child in the Bay of a Thousand Islands...about two miles above Namu, one of the haunts of the *boq's*. He heard a number of the creatures in the forest behind him and seized his gun, at the same time calling out to them to go away. Instead the breaking of branches and beating upon tree-trunks came nearer. Becoming alarmed he called out once more: 'Go away, or you shall feel my power.'

"They still approached and *Qak'lis* fired in the direction of the sounds. There followed a wild commotion in the forest, roars, grunts, pounding and the breaking of branches. The hunter, now thoroughly alarmed, told his wife and child to embark in the canoe while he covered their retreat with his gun. He followed them without molestation, and anchored his craft not far from shore. The *boq's* could be heard plainly as they rushed to and fro on the beach, but only the vague outline of their forms were visible in the darkness...[The Native then describes how the mountains appeared higher to him and found himself in shallower water, which he attributed to some supernatural power of the *boq's*.] *Qak'lis* jumped overboard into water which reached only to his knees, and towed his canoe to Restoration Bay, the *boq's* following him along the shore."

McIlwraith continues: "This is not the only the only occasion on which *boq's* have appeared near Restoration Bay. Within the lifetime of the father of an informant, a chief set out with some friends from Kwatna bound for Namu. They traveled overland to Restoration Bay, thence by canoe, making the journey without incident. While returning, they decided to gather clams on the rocky point of the bay. As the canoe shot around the tip of the promontory, they saw a *boq's* gathering shellfish. The paddlers backed their canoe behind some rocks whence they could watch without being seen. The creature acted as if frightened; it kept looking backwards, then hurriedly scraped up some clams with its forepaws, dashed off with these into the forest and came back for more. The chief decided to attack the animal. A frontal approach was impossible owing to lack of cover, so he landed and crept steadily through the forest, armed with his Hudson's Bay Company's musket. Presently he stumbled upon a heap of clams which the animal had collected. He waited until it returned with another load, then raised his musket and fired. Instead of killing the *boq's,* its supernatural power was so great that the hunter's musket burst in his hands, though he himself was not injured. The *boq's* shrieked and whistled as if in anger, and at once hordes of its mates came dashing through the forest. The frightened chief rushed out on the beach and called to his comrades to save him. They brought the canoe close to shore so that he could clamber aboard, and then paddled away unharmed."

McIlwraith adds, "The Bella Coola believe that the *boq's*, unlike most supernatural animals have not abandoned the country since the coming of the white man. One man was most insistent that they still lived on King Island, and promised to point one out if a visit were made to that spot. This man refuses to camp at the place where, he affirmed, *boq's* are common. Another informant stated that though he had never seen one of the monsters, a horde of them surrounded his camp near Canoe Crossing for a week. Every night he heard them roaring and beating upon trees and branches." [These descriptions accord closely with modern-day accounts of sasquatches seen foraging for shellfish, taking the impact of medium-caliber bullets without going down and exhibiting aggressive vocal displays when angered by man. For more on the possible ability of the sasquatch to withstand high energy bullets see the chapters on sightings involving Mr. Steve B., pages 47 and 226]

Concluding, the ethnographer notes: "*Boq's* have been heard as recently as 1924 according to popular belief. In January of that year a number of young Bella Coola were returning home in a motor boat from Ocean Falls. They camped for the night on Burke Channel, and were alarmed to hear a crashing of bushes and a beating on tree-trunks. Thoroughly frightened, they directed the beams of several electric torches [flashlights] in the direction of the sounds without avail and at last started the engine of their motor-boat, the noise of which frightened the *boq's* away."

Regarding the similar-sounding *atlakwis,* McIlwraith says, "An animal somewhat similar to the *boq's* is the *atlakwis*. The Bella Coola state that long ago these lived as human beings near Port John on Evans Arm, King Island. They were a strange people, with peculiar rites including the worship of fire. Their neighbors, the Bella Bella, from whom they differed entirely, determined to destroy the *atlakwis*. A massacre took place, and only a few survivors escaped to an island in one of the lakes in the interior of King Island. Here in the course of generations their bodies grew hairy and lost their human characteristics; it is said that a few paintings made by them are still visible there. Not many years ago some Bella Coola saw what may have been the last surviving *atlakwis*. Two hairy creatures were observed fishing from a skin boat on Burke Channel. As soon as they realized that they were observed, they dashed ashore, broke up the framework of their canoe, and plunged into the forest carrying its skin covering. Nothing more seems to be known concerning them, nor do they appear ever to have granted supernatural assistance to anyone."

This is a strange narration. If based on events as related, it may be faintly suggestive of a relic population of *Homo erectus,* not thought to have survived more recently than 250,000 years, but credited with fire and simple tools. Although not matching the typical sasquatch type, creatures of similar but man-sized or smaller form are very occasionally mentioned in Alaskan folklore and sighting reports.

Heiltsuk Beliefs

The Bella Bella (Heiltsuk-speaking) people inhabit the village of Bella Bella and the archipelago of British Columbia islands between the Tsimshean to the north and the Kwakwalawala (Kwakiutl) to the south. Among the Heiltsuk, the creature most resembling the sasquatch is recorded as *pkw's* and is represented in masks of great style and ornamentation. Wide, flaring nostrils, projecting brow ridges often ornamented with fur, deep-set eyes and grinning rows of even teeth are traits of these masks. Folklore surrounding the *pkw's* in Bella Bella is mixed with stories of the cannibal giantess *k!a'waq!a*, similar to the Kwakiutl *dsonoqua*. Boas recounts two tales of this creatures.

"The *K!a'waq!a* [*dsonoqua*] carries a basket on her back in which she carries away children. Once upon a time she [he?] stole from the village *Xune!s* a boy whom she kept in her house [variously translated as any form of shelter, including a hut or lean-to] for a whole year. One day *ask:a'naqs* [person's name] came to him and whispered into his ear, 'The *K!a'waq!a* goes always down to the beach to get mussels and clams. When she comes up the steep mountainside, put the siphons of the clams on the tips of your fingers and move them towards her, opening and closing your fingers.' The boy obeyed. When the *K!a'waq!a* came up he moved his fingers and shouted, '*ananai'dzedzeq!ei' K!a'waq!ai !*' She became frightened and asked him to stop. When he continued she fell down the cliff and was dead."

The following second account is similar in theme.

"In *Hau'yat* where people lived in summer, children disappeared one after another at night time. The people did not know what happened to them. They asked four courageous men to watch all night. The watchers armed themselves with clubs, spears and lassoes. At midnight the *K!a'waq!a* appeared. They threw a lasso around her neck, speared her and cut off her head. The next morning the people deliberated what to do with her body. Some wanted to tie stones to it and throw it into the sea. Others wanted to burn it. Finally all agreed to burn it to ashes. Those who breathed in the smoke of the fire died of it. The others took to their canoes and left the place."[7]

While these narratives are imbued with mythological motifs,

they do also mention recurring traits such as shellfish foraging, abduction of children and give a possible allusion to the Native killing of some kind of very strong manlike creatures.

Kwakiutl Beliefs

Midway down the British Columbia coast the Kwakiutl (now Kwaguilth) peoples traditionally speak of *dzonoq!wa* or *dsonoqua*, a female hair-covered giant, who inhabits the coastal forests, is nocturnal, whistles to communicate and in ethnographic narratives, is fond of abducting children. Her dark face is dominated by projecting brow ridges, closed sunken eyes and lips pursed in a very apelike manner. Of this entity, the famous ethnographer Franz Boas wrote: "The *Dzonoq!wa* are a people who dwell inland or live on mountains. Their houses are far in the woods. One of them lives in a deep lake far on top of a mountain. A trail leads back from the human village to a pond near which their houses stand [may translate "shelters are located"]. The *Dzonoq!wa* have black bodies; eyes wide open, but set so deep in the head that they cannot see well. They are twice the size of a man. They are described as giants, and as stout. Their hands are hairy. Generally, the *Dzonoq!wa* who appear in the story is a female. She has large hanging breasts. She is so strong that she can tear down large trees. The *Dzonoq!wa* can travel underground [dig or use caves?]. When speaking they [*dzonoq!wa*] pronounce the words in such a way that every syllable of ordinary speech is repeated with initial "h" substituted for the consonantal beginning of the syllable, or with "h" introduced before the initial vowel. Their voice is so loud it makes the roof-boards shake."

Boas notes several miscellaneous transformational beliefs about these strange people of the mountains and then adds several more mundane attributes: "When she has stolen a child she keeps it as her daughter and picks salmon berries for her." Also: "She is in the habit of going to the villages to steal salmon. Then she throws aside the roof boards and reaches down to take the fish from the drying frames." (Boas, 1932, pp. 142–5.)

A similar figure from Kwakiutl folklore is *buk'wus* or "wild man." *Buk'wus* masks are still prevalent in the works of Native carvers and usually portray the same deep-set eyes, although usually open, the pronounced brow ridge, and a profusion of hair. Of *buk'wus*, or "woodman," Boas notes (p. 146): "The woodman is a

Figure 24:
Photograph of totem representing Kwakiutl *dsonoqua* holding her baby, on display at the B.C. Provincial Museum, Victoria.
Photo: David Hancock

person who takes away drowned people. His body is as cold as ice. Whoever accepts the food offered by him cannot go back. To those who are taken away by him his house [also translated "dwelling"] appears in the night, while as soon as day breaks, it disappears." There are other supernatural powers attributed to the "woodman" and Boas states: "This being evidently corresponds to the land-otter spirits of the northern tribes." As noted by Rohner and Rohner,[8] *buk'wus* "is generally considered to be harmless and has the peculiar ability of moving himself instantly from one location to another a great distance away."

In some narratives by other ethnographers, he is attributed with the habit of striking trees with a piece of wood. Although *buk'wus* is man-sized, or occasionally smaller, he can leap through the forest with a stride four times that of a normal man. His body is hairy, he is shy of man and has a frightening countenance. Those lost or drowned may, in some stories, be enticed into his world by eating his food or being carried off bodily. Occasionally those so lost may be brought back to civilization through purification similar to rites practiced in former times by the Tlingit and Haida. Today, *buk'wus* continues to be spoken of by members of Kwakiutl communities.

Figure 25: Photograph of mask representing Kwakiutl *buk'wus* or "wild man of the woods."
Photo: Author's collection

Comox Beliefs

The Comox occupy the east half of Vancouver Island around Campbell River from Courtenay in the south to Sayward in the north, and on the mainland around Toba and Bute Inlets. The Comox language group includes Sliammon at Powell River, on the B.C. mainland.

Although I have no specific ethnographic references, much popular Vancouver Island folklore surrounds the naming of Forbidden Plateau, a large subalpine area on the east slopes of the Vancouver Island Range behind Comox. In brief, Natives of the Comox group historically told of a time in the 1800s when a large party of Comox was traveling through the plateau with their families. The men were caused to move away suddenly to defend against an incoming attack by another tribal group, and obliged the women and children to stay at their camp in the plateau to await their return. Approximately forty women and children were said to be involved. Upon the return of the men a short interval later, they discovered the entire body of women and families gone. No trace was ever found. Persistent tribal belief was that the women and children had all been abducted by a race of "mountain giants," and the area became forbidden to those who returned to the coast. Now an attractive alpine recreation destination for hikers, skiers and hunters, it has retained its name Forbidden Plateau.

In 1976, one Native Comox elder, of the mainland Sliammon tribal group near Powell River, told me that the Comox word for a wild, hair-covered manlike creature was *mai-a-tlatl* and that belief in such creatures had been widespread among his peoples. He was unable to give any more precise details of such creatures, only that they "lived in the forest" and were generally man-sized.

Nootka Beliefs

The Nootka traditionally inhabited much of the west coast of Vancouver Island; they are now known as Nuu-chah-nulth. Besides Nuu-chah-nulth variations of *buk'wus*-like creatures, there exists a traditional belief in the existence of a monstrous hairy giant referred to as *matlox*. In 1792, Jose Mariano Mozina, a botanist and naturalist accompanied Bodega y Quadra on his voyage of exploration from Mexico along the coast of present-day British Columbia. He conducted a survey of Nuu-chah-nulth lifestyle, known animals, ethnobotany and language for the Spanish Viceroy of Mexico. Following is an excerpt from one of several published accounts of *Noticas de Nutka* [Nootka Notes].

"One does not know what to say about a Matlox, inhabitant of the mountainous country, of whom all have an unspeakable terror. They figure that it has a monstrous body, all covered with black animal hair; the head like a human; but the eye teeth very large sharp and strong, like those of the bear; the arms very large and the toes and the fingers armed with large curved nails. His howls fell to the ground those who hear them, and he smashes into a thousand pieces the unfortunate on whom a blow of his hand falls."[9]

In this translation afforded me by John Green, there is special emphasis placed on the attribute that would clearly distinguish the anatomy of a bear from that of a primate. In the text edited and translated by Iris Higbie Wilson[10] of the University of Washington, there is less differentiation made between characteristics which could be translated loosely as belonging to a bear or a hominoid. However, her emphasis remains on manlike characteristics, and it is not likely that the Nootka, who hunted and utilized bears as a resource, would fall into any confusion between the two.

The name *matlox* has phonological similarity to the preceding Comox name *mai-a-tlatl*, and this suggests one or the other may be a loan word, although in which direction is unclear.

Another Nuu-chah-nulth being who was said to inhabit wild places is *pokmis*. Sapir and Swadesh[11] noted that *pokmis* were regarded as wild, hair-covered creatures who had originally been human but, as a result of becoming immersed at sea or through exposure in the forest, became transformed. If the word *pokmis* receives a consonant shift from "m" to "w" becoming "*pokwis*," it resembles one more northern pronunciation "*puk'wis*." Fortunately, linguists have solved the confusion for so many similar sounding names for the "wild-man-of-the woods." The Kwakiutl *buk'wus* is considered to be the oldest form, providing the basis of the loan words to other languages north and west. Currently, the ter *"cacuuqhsta"* is used by the Nuu-chah-nulth to refer to sasquatch-like creatures.

Coast Salish (Nanaimo) Beliefs

The Nanaimo Native group traditionally occupied the portion of eastern Vancouver Island surrounding Namaimo and its offshore islands. In Nanaimo, B.C., located near the related Cowichan on east-central Vancouver Island, one Native elder of the Nanaimo tribal group gave me the following information about names of creatures in the Nanaimo dialect of Halkomelem Salish.

"There are three types of creatures similar to sasquatch. The first two are identical to the mainland creature, are hairy and black, a bit larger than a man and called *squee'noos* and *papay'oos*. Even a white man could see these creatures and they have [had] no special powers.

"However, you could get one as a guardian spirit and it would make you really strong, but they would also make you a little bit, well, unlucky. I knew a fellow once who had one as a guardian spirit and it made him really strong. He could handle the peavey pole [on log booms] with just one hand. But nothing ever seemed to go quite right for him. That was the only problem with having one as a guardian spirit.

"The other creature was *kwai-a-tlatl*—the 'tree striker.' It was a lot like the other two but would knock down trees and make a big sound. If you ever tried to follow their tracks they would lead you around in a big circle and make you go crazy.

"My dad and I had been fishing at Dodd's Narrows, south of Nanaimo, in the forties and we had heard a tree fall just onshore in the forest. I asked my dad, 'What was that?' My dad replied. 'That was a *kwai-a-tlat*. Don't even think about going over there.' But late that night I did sneak off the boat in the skiff and rowed to shore. I tried to find the tree, but couldn't, and came back before long. The next morning on the boat, my dad seemed to know I had gone looking; he said, 'You went, didn't you?' and seemed a little mad at me but that was all."[12]

Another aspect of belief that was generally widespread among the Coast Salish, but never much spoken of, was the belief that abduction of women by sasquatches did in fact occur on rare occasions. An elderly Nanaimo woman with whom I was acquainted, Mrs. M. B., told me in 1974 that her people had stories, which she believed, of women carried off by *squee'noos* around the turn of the century. In one case the woman was her grandmother, and after either returning from the forest, or being returned, she gave birth. The infant was sadly malformed and was in fact stillborn. As she told me, "The women of my family helped my grandmother let the baby's body slip off the ferry, out in the middle, halfway to Vancouver.

"That was not the only time this happened. Another woman was stolen and lived with the *squee'noos* in the mountains for more than a year. This was around the same time. She brought back with her a very strange baby, which lived for a little while. He was not right and died soon after their return. I don't think he was even a year old. These things happened in my grandmother's time, but nothing like that has happened for a long time. We don't like to talk much about them."

These accounts may, of course, have been based on the return of women to the community following an unsanctioned affair or elopement, the proof of which needed an explanation. However, this was traditionally unremarkable, if not somewhat commonplace, and Native society was generally supportive of the mother within her brother's family, moiety or clan. Even with the advent of missionaries, such a woman was not overly censured. For perhaps this reason, such narratives may be worth some attention. That half-sasquatch (the man-sized *squee'noos*) offspring existed, at least briefly at one time, was certain to her.

Coast Salish (Musqueam) Beliefs

The late Professor Wayne Suttles,[13] an authority on Coast Salish ethnography, noted how the Musqueam, who live at the mouth of the North Arm of the Fraser River, at what is now Vancouver, B.C., spoke of a species of manlike animal. The male was called *sesq'ec* (sasquatch) and the female *qelqelitl*.

"It must have been long ago when there were still only Indians here and everywhere on up the river too there were only Indians. There were none of those who are called "white people" but only Indian people. According to the old people, walking in the woods everywhere away from the water were what are called the *sesq'ec*. They were big, resembling a tall person but tall, far taller than the biggest people here. And it is said that their wives were what the people called the *qelqelitl*."

Coast Salish (Upper Stalo) Beliefs

The classic term sasquatch is derived from a Coast Salish language (Halkomelem) from the lower Fraser Valley; *suhsq'uhtch* is perhaps easier to pronounce correctly than a phonetic spelling. For Natives from Chilliwack upstream on the Fraser River and south to the

American border, British Columbian ethnographer Wilson Duff[14] used the ethnic term "Upper Stalo." He wrote of two types of creatures described by Upper Stalo informants. One was a cannibal woman, but the other was a classic account. "Sasquatches are usually seen singly. They are described as men, covered with dark fur, more than eight feet tall, who leave footprints about twenty inches long."

He provides two older accounts and two newer ones. In an older account of typical form narrated by Adeline Lorenzetto of Ohamil, he lists the following attributes: "The sasquatches would cause unconsciousness if they touched a person, they would abduct women whom they would keep, they would cause those women to have half-human children and would steal fish and other food for their Native wives and the children. They were said to have some form of simple language, which some women learned and if the woman managed to escape and reenter society, she would suffer bouts of unconsciousness because she had been with the sasquatches and wasn't like a human anymore, she had forgotten her language and "hair was starting to grow all over her body."

In one story, Indian doctors worked on a woman and she became normal again. Many years afterward the sasquatches came back, but the woman could no longer speak their language. However she asked that none of them be harmed. Some might be relatives of hers. In the other older narrative a sasquatch killed a group of women but left their children unharmed.

In narratives of encounters with sasquatches dating from the 1990s, Duff stated that a person would see a sasquatch, usually on a moonlit night, run from it and would be followed but left unhurt to escape. One account related the shooting of a sasquatch and in the other, an event documented historically, a house at Ruby Creek was broken into by a sasquatch (for dried fish) which damaged the house and the fish barrels.

Coast Salish (Lummi) Beliefs

Suttles[15] interviewed and wrote the accounts of two men who described sasquatches in the Lummi area, north of Bellingham.

"The *c'amek'wes* is a great tall animal or whatever it was that lived in the mountains. It was like a man but shaggy like a bear, like a big monkey 7 feet tall. They went away when the Whites came.

(The Indians never killed any; it was a pretty wise animal, or whatever you call it.) If you saw one it made you kind of crazy. They throw their power toward you.

"Over forty years ago some fellow across the line went hunting deer early one morning when snow was on the ground. He saw one [a *c'amek'wes*] and followed it to the edge of a lake where it had disappeared. He went home and got kind of crazy. His wife put him to sleep by the fire (they were living in a kind of smokehouse) and while she was out getting wood he rolled into the fire and died." (Suttles' interview with Julius Charles.)

"The *c'amek'wes* are big, 7 to 8 feet tall. They whistle only, can't talk. They whistle when you go out in the evening. Once some White people caught one and tried to feed him. They gave him potatoes. He picked them up, looked at them and threw them away. They gave him meat and he did the same thing. I guess some make you crazy. They are real *stlaliqem* [meaning powerful]. They grow hair on the body. There are none here anymore, but I guess there are some up in the mountains around Chilliwack. If a person could get one for [Suttles merely states word for power/guardian spirit] I guess it would be pretty tough. (No I never heard of one with it. I don't know what they eat.)" (Suttles' interview with Patrick George.)

Coast Salish (Twana) Beliefs

The shores of Hood Canal and the Skokomish River watershed on the southwest shores of Puget Sound, Washington, are the traditional home of the Twana peoples. W. W. Elmendorf[16] worked with them in the 1930s and recorded the following beliefs.

"Mountain and forest giants (*c'iatqo*) were generally referred to in English as stick Indians, the Chinook jargon term *stik* meaning 'forest.' These creatures were of human form, taller than normal human beings, lived in the mountains or rough foothill forests, went naked except for a breech clout, had odorless bodies, which enabled them to walk up to game and kill it before the animal scented them, and could climb vertical cliffs and leap great distances. They were usually invisible. People feared the *c'iatqo* but seem to have suffered little harm from them beyond occasional thefts of killed game. Henry Allen, a Twana informant, had heard they 'could make people crazy' but did not know how this was done. They did not function as soul stealers."

These attributes generally match those of sasquatches if we ignore the notion of wearing a breech clout (it is not clear why a presumably hair-covered creature would have need of such limited clothing) and if invisibility might be equated with "disappear quickly," such as all the great apes except the gorilla are quite capable of doing.

Cross-Cultural Comparisons

This collection of narratives leaves us with the impression of a creature or perhaps several types of similar hair-covered creatures, generally avoiding man and inhabiting the coast of the Pacific Northwest. Separate names and attributes for male and female creatures are not necessarily significant, as anatomy and behavior for males might well differ from that of females, and creatures in the folklore of lesser size may have had an original basis in real sightings of juvenile or adolescent creatures. There is no denying the various functions that stories of sasquatchlike creatures have had in traditional Native culture. The many recurring themes such as transformation (*kushtakaa, gagiit*), capturing the drowned (*kushtakaa, gagiit* and some *buk'wes*), and stealing and/or cannibalizing children (*dsonoqua* and similar giantesses) may be viewed as recurring folk motifs with regulatory functions in culture.

It is also fair to say that while anthropologists have been very successful in suggesting a number of cultural reasons for "inventing" the existence of wild, hair-covered creatures resembling the sasquatch of modern description, they cannot exclude the possibility that a widespread belief in such creatures may have actually had some basis in reality. Professor Wayne Suttles examined the Coast Salish beliefs in detail and concluded that only several of all the Puget Sound entities qualify as possible candidates for sasquatches, many of them including characteristics that would place them in a category more suited to elements of tribes from the interior, who had at times made raids upon the Coast Salish.

There persists, among some southeast Alaskan locals, the notion that most modern southeast Alaskan reports of sasquatches describe man-sized creatures and that this is somehow related to the description of the hair-covered *kushtakaa* of Tlingit folk belief. While the latter is an undisputed fact of ethnographic anthropology, the thought that sasquatches are man-sized is entirely unsupported by the database of reports. Southeast Alaskans of the older generation are quite familiar with the term *kushtakaa*, or wild man, but do not usually associate *kushtakaa* stories with sasquatches. For them, the *kushtakaa* has definite connotations with Native belief in the supernatural. Many younger residents are of families emigrated from "the lower forty-eight" states and are not familiar with the term *kushtakaa* at all. Since silence surrounds the telling of the older stories, there is an aura of disbelief, ridicule and occasionally humor regarding any recent reports of sasquatchlike creatures in the vast archipelago that makes up southeast Alaska. Overall, reported heights present a bell-shaped curve of statures, with a mean height of something close to seven feet. This is exactly what one would expect with differences in age and gender of the creatures, and the lesser frequency of five-foot and also eight—or nine—foot reports seems to correspond well with projected heights of juvenile and adult male sasquatches.

In summation, the varied ethnographic material surveyed here may also be viewed as generally supporting the notion of a sparse population of secretive, nocturnal, manlike creatures inhabiting the mountains and forests of the coastal rain forests. Given the common theme of sasquatchlike creatures spanning the entire coast from Yakutat, Alaska, to Puget Sound (and southward, beyond the scope of this book, to the central coast of California), the persistence of such beliefs in many areas long past the arrival of non-Natives that "a real animal is still there" seems to be one very plausible explanation. If one asks, "Is there any traditional Native ethnic group from Yakutat to Puget Sound that does not have traditional 'knowledge' of such an animal?" then the answer, on the basis of the preceding ethnographic accounts alone, would have to be, "None."

Following is a cross-cultural comparison of traits most often attributed to sasquatchlike creatures by various Native groups. Traits listed include giant size, living in the mountains, nocturnal, hair-covered, whistle, cry or shout as a form of communication, steal food or people, to have been killed by humans and finally transformation, i.e., the ability to shape-shift into another form such as that of an otter or a human.

The presence of a given sasquatch trait in the folklore of listed Native groups is indicated by the sign "+" and its absence is indicated by "-". Where the presence is unknown the symbol "?" is

Table 1: Cross-Cultural Comparison of Sasquatch Traits

Tribal Group	Tlingit	Haida	Tsimshean	Bella Coola	Bella Bella	Kwak.	Kwak.	Nootka	Comox	Nanaimo	Stalo	Puget Sound	Lummi	Twana
Being Trait	*kushta-kaa.*	*gagiit*	*baoosh*	*bukw's*	*kawaqa*	*dsonoqua*	*bukwus*	*matlox*	*mai-a-t latl*	*skwee-noos*	*ses'qec*	*cyatkwu*	*camukws*	*ciatqo*
giant size	-	-	-	-	+	+	-	+	-	slightly	+	+	+	+
in woods/mtns	+	+	+	+	+	+	+	+	+	+	+	+	+	+
hairy	+	+	+	+	+	+	+	+	+	+	+	+	+	+
fast	+	+	+	+			+				+			+
whistle	+	+	+	+	+	+	+	roar				+	+	
speak	rare	rare	-		slight	slight					+		-	
steal food	+		+	+	+	+					+	+		+
steal women										+	+	+		
have half-human child										+	+			
steal children				+	+							+		
cause unconsciousness	+	+									+			
steal souls	+	+				+			-				+	+
kill people	-	-	-	+	+		+		-	+				
trick people	+	+	+			+						+		

used. Other traits less commonly mentioned have been deleted for the sake of brevity.

For the Coast Salish groups south of Nanaimo, data is condensed from a similar table for Puget Sound and the lower Fraser Valley by Dr. Wayne Suttles, included in the second chapter of his book *Coast Salish Essays*.[17] In examining the relative frequency of the traits in Table 1, we find uniform traditional belief in the existence of a continuous population of hairy creatures, usually of giant size, all up and down the coast, except in southeast Alaska where traditional belief holds them to be more man-sized. Alaskan beliefs are also uniform on the subject of being able to transform back and forth between "land otter" (a river otter) shape and human form. The trait of whistling, possibly a form of communication, is mentioned in twelve of eighteen ethnographies. It may have been that the whistling of an otter was at one time likened to the vocalizations of sasquatchlike creatures. The ability to swim and live off of the intertidal zone are important similarities between otters and earlier reports of sasquatchlike creatures, as described in traditional beliefs. It is also possible that either early Alaskan beliefs were based on observation of actual adolescent and subadult creatures following a person gone missing or, alternatively, that there existed a smaller northern race of the larger southern form of sasquatch, possibly not the same exact species. Modern sasquatch reports from Alaska do seem to contain a somewhat higher frequency of man-sized, hair-covered creatures, but such reports are still less common than the seven- or eight-foot variety. The stealing of food is mentioned in thirteen of eighteen entries. The abduction of humans is mentioned in half of all the above ethnographic sources but, among the three Alaskan Native groups, so is the alleged killing of humans.

South of Alaska the pattern is somewhat more uniform, with nocturnal activity, stealing food and whistling mentioned the most frequently. Abduction of humans is often mentioned. This trait seems to be confined just to the northern half of the regions studied and if the folklore concerning northern creatures also includes allusions to a second smaller-sized type of hominoid cohabiting with the conventionally reported sasquatches, then perhaps it is more aggressive as well. This is admittedly speculative. Such tales can be found in many Native ethnographies, and their absence from much of the work among the Coast Salish may be simply due to a stronger original motif among the more northern Native stories. That the Tsimshean *ba'oosh* appears to be a word borrowed from more southerly neighbors, is generally accepted among West Coast linguists.

Are all of the traditional beliefs possibly describing simply humans "gone wild"? It appears not. Some of the traits that do evidence themselves most uniformly, such as "hairy," have often been said by skeptics of sasquatches to represent characteristics of humans from neighboring hostile tribal groups. That is to say, sasquatches are merely caricatures in the folklore representing historically hostile interior tribes. According to such a view, these characteristics of sasquatches represent all that is inhuman or undesirable in society. This has been dragged out by all cautious authors, including the most respected academics. To my less-lettered mind, "hairy" is simply not a characteristic of any Pacific Northwest Native group. Perhaps some individuals from a hostile tribal group would allow their hair to grow long, but long hair was also universal among shamans and others in most all coastal cultures, as were thin moustaches and sparse beards among some Native men. Simply not shaving such scant facial hair with an abalone razor does not bring to mind the appellation "hairy," let alone the term "hair-covered." If we view "hairy" in the bodily sense, not as merely "unshaven" or "unkempt" about the face, it is apparent that there is a profusion of body hair that distinguishes the sasquatch from a human "gone wild." There is no other characteristic more commonly reported for sasquatches, either in folklore or current reports.

There has never been a single case where any lost human, mentally defective or otherwise, has been able to gain a foot in stature, develop broader shoulders, change the shape of his foot so that the heels become broader or grow a coat of hair on the body. Of course, a long head of hair and a beard are possible, but the body becomes only tanned, not hairy, and the tendency is to become lean not heavily muscled. This is not to say that lost people cannot survive to be later identified by some as *kushtakaas* or *gagiits* or that such a belief is not extremely therapeutic in obtaining the closure needed in cases of long-missing relatives.

Exactly how the early Native carvers pictured the physical appearance of sasquatchlike creatures is more varied and subject not only to the oral traditions of Native groups, but also somewhat to the idiosyncrasies of individual carvers. In many cases it is likely that the northwest stylized form of art contributed to certain physical traits being portrayed. It is also noted that today, many Native artists choose to incorporate nontraditional features into their masks and totems. Reference to these newer works of art has been avoided in order to help define traditional interpretations.

Examining regional dissimilarities as a possible function of real animals' variations may have value in reported sightings of such animals, but its relevance to the slight differences between sasquatches in Native ethnographies is questionable. There are simply too many variables affecting the origin and historical oral transmission of belief systems to make specific conclusions about how such hypothetically "real" creatures might differ from region to region. We can make some observations based on consensus, however. The suggestion is clearly made that "Native" sasquatches are generally agreed upon as being large, hairy, thought to live in the mountains, whistle as at least one form of communication, are nocturnal and are often given to stealing food. What all this may suggest to those who wish to interpret any one ethnographic sasquatch as being descriptive of a real primate is open to conjecture, but the implications are obvious enough.

It may be wise to note that it is possible for a dozen different ethnic First Nations to all agree quite closely on the physical attributes of any animal species, perhaps even a number of ascribed behaviors, and yet still hold a belief in any number of different metaphysical attributes for that animal. It is common to hear traditional Natives say that while their tribal areas have, for example, a wild man that "carries a stick and cries like a baby," that on the other side of the inlet, that other tribal group believes in a wild man that is different because it "steals women and strikes trees." They may even point out that their own version of a generic sasquatch differs substantially from the similar creatures of neighboring ethnic groups on the basis of these supernatural abilities alone. Giving different names to such a creature within ethnic and political boundaries, however, does not automatically make those animals different species in the eyes of zoologists. But traditional Natives were apparently perfectly comfortable with the notion. It may also be that some individual attributes of sasquatch-type creatures were only "known" to one ethnic group, but that over many tribal areas, the collective individualities of sasquatches could be pieced into one larger "mosaic" of characteristics that might present a complete profile to us.

That we have better knowledge of the forests and the mountains of the Pacific Northwest than the Native inhabitants of this vast area seems open to some dispute. If one argues that the Native folk tales all describe a similar creature for some strange reason of cultural sharing from California to Alaska and beyond, then one is also hard put to explain how we non-Natives have also "happened" to have come up with our own similar version of the creature. Borrowing a belief in earlier times from Native cultures was not very common unless there was a very practical reason for it, such as knowledge related to the physical environment or names of animals likely to be encountered in it. I suggest this has been the case for the sasquatch.

1. Swanton, John, 1909, *Tlingit Myths and Texts*, p. 48.
2. McNair, Peter, et al, 1999, *Down from a Shimmering Sky*, University of Washington Press, Seattle, p. 169.
3. Brown, Albert, 1988, "Alaskan Bigfoot," *New Alaskan*, October, 1988, pp. 14, 15.
4. Dunn, John A., 1995, *Sm'Algyax: A Dictionary and Grammar for the Coast Tsimshean*, University of Washington Press, Seattle.
5. Barbeau, Marius, 1929, *Totem Poles of British Columbia*, Nat. Mus. of Canada Bulletin 61, Ottawa, p. 168.
6. McIlwraith, T. W., 1942, *The Bella Coola Indians*, University of Toronto Press, Toronto, Vol. I, pp. 60–63.
7. Boas, Franz, 1932, *Bella Bella Tales*, The American Folklore Society, New York, pp. 96–97.
8. Rohner, R. P. and E. C. Rohner, 1970, *The Kwakiutl—Indians of British Columbia*, Holt, Rinehart and Winston, NY, p. 67.
9. Green, John, 1978, *Sasquatch: The Apes Among Us*, Hancock House, Saanichton, B.C., p. 25.
10. Mozina, Jose, 1970, *Noticas de Nutka,* ed. Iris H. Wilson, Univ. of Washington Press, Seattle.
11. Sapir, E. A. and M. Swadesh, 1939, *Nootka Texts: Tales and Ethnological Narratives*, Linguistic Society of America, University of Pennsylvania, Philadelphia.
12. Alley, Robert, unpublished papers.
13. Suttles, Wayne, 1987, "On the Cultural Track of the Sasquatch," *Coast Salish Essays,* University of Washington Press, Seattle, p. 80.
14. Duff, Wilson, 1952, *The Upper Stalo Indians, B.C. Provincial Museum*, Anthropology in British Columbia. pp. 118–19.
15. Suttles, Wayne,1987, p. 82.
16. Elmendorf, W. W., 1960, *The Structure of Twana Culture*, Washington State University Research Studies, Monograph Supplement 2, pp. 532–4.
17. Suttles, 1987, p. 99.

Part Two / Circumstantial Evidence

11 Big Footprints in the Alaskan Panhandle

"I kept moving quickly along the deer trail and a few moments later, while still scanning for deer tracks, I saw a footprint. I had stepped right over it, stopped, realized what it was and took a good look. It was about sixteen inches long, six inches at the ball, and about four, maybe four and a half, inches at the heel. The heel was definitely narrower. Most of the track was about three-quarters of an inch deep, parts were one to two inches deep. It was set in mud and clearly showed five toes. The toes were about two and a half to three inches long. It had a kind of hourglass shape as thought the heel and ball had imprinted more heavily. There were no claw marks..."
M. A., Ketchikan, Alaska, 1993

One question often heard regarding sasquatches is, "Why is there no evidence?" Some anthropologists, biologists and authors have pointed out that there is, and it's right at our feet. Tracks, in fact, are reported almost everywhere and tell quite a good deal about who or what made them. They can be subjected to scientific scrutiny, cast, measured and, through the study of dermatoglyphics, evaluated forensically.

The anatomy of the sasquatch foot, as analyzed from hundreds of sets of tracks in the Pacific Northwest by Krantz[1] and Meldrum,[2] is now well documented. They are larger than bear tracks and much wider at the heel. A human track might resemble a hypothetical juvenile or adolescent sasquatch track but has a much narrower heel and shows a narrowing for the longitudinal arch. While a juvenile sasquatch may leave a track of only ten inches, the first toe is still larger than the rest, unlike those of a bear, whose largest toe is the middle and whose smallest is the innermost—what we would call our "big toe." Brown bear or grizzly tracks leave claw marks even

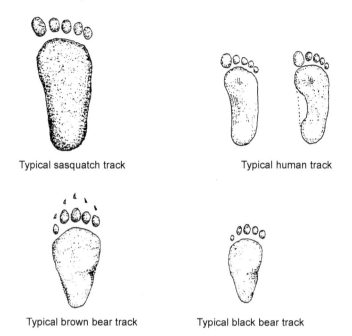

Figure 26: Comparison of human, sasquatch, brown and black bear tracks, to scale.

when front and hind feet register on top of each other. There is really no similarity.

The average size of reported sasquatch tracks is close to the figures given by Sheldon and Hartson[3] in their field guide *Animal Tracks of Alaska*, i.e., fourteen to seventeen inches in length with a width to seven inches. Occasionally there have been documented tracks, measured and cast, in the order of twenty-plus inches with width up to eight inches or more. These tracks may be ostensibly assigned to unusually large males. It is important to remember when considering such outlandishly large tracks that hominoids such as *Gigantopithecus*, if upright as advocated by Krantz's "mandibular flaring" hypothesis, could possibly have been as tall as the twelve feet postulated by early Chinese and Russian paleontologists. Since we are being conservative, we will adopt Simon's[4] early estimate of eight feet for *Gigantopithecus* when standing upright.

The stride of any two-legged tracks is measured from the toes of one right print to the toes of the next right, or from one left to the next left, in other words, two steps, not one. The stride in reported

Figure 27: Photograph of cast of a 15 ½-inch track recorded by Dr. John Bindernagel on Vancouver Island, B.C., 1998. [T88A-map11]

sasquatch tracks is usually around six feet, but in some track reports (running) it reaches up to twenty feet. The distance between right and left heels is usually two to four feet, with commonly reported distances of three to three and a half feet.

Sasquatch tracks are often reported as deposited overnight and close to large bodies of saltwater, but also by streams, lakes, muskegs and even logging roads. Beach sand or mud has yielded more tracks along the Alaskan and British Columbia coasts than other surfaces, but tracks in snow follow a close second. The following chronology details track reports from southeast Alaska that have come to be documented and are suggestive of a large, biped primate.

A retired Haines fisherman, Mr. Thomas Abbott, reported in Ketchikan in 1998, that in April of 1974 he had found an unusual track on a beach north of his town. Haines is north of Juneau on the Lynn Canal.

"It was April, 1974, I had just got married that year, and in April I had gone looking for seals with my partner in the small inlet about ten miles north of Haines, up Taiyasanka Inlet. Up the far side of the small inlet we stopped to examine something strange on a small, flat promontory covered with a couple of inches of snow. The snow had been trampled down flat and there were dozens of wolf tracks and blood all over. The flat snowy area was between the beach and the forest and was about forty by thirty feet or so. What was most remarkable was a single, large, eighteen-inch track showing toes and a wide foot. It was not a bear. The track was quite different. Also in the flat area we found a handful of light brown hair, about the same length as the track. I've often since wished I'd had a camera or just kept some of that hair. There's lots of stories you hear about those things up there. My dad told me about seeing a big one one time out of Hoonah, when he was a kid. It might seem strange to hear, but it's no big deal up there." [T74A-map1]

Something leaving two-legged tracks on top of a sharp, snow-covered ridge caused pilot Chuck Trayler of Wrangell and Stan Hewitt of Ketchikan to circle in their plane twenty times just 100 feet above the tracks near Thom's Creek, Wrangell Island, in early 1980. A brief description of the report, with a photo, appeared in the Wrangell *Sentinel*, Feb. 20, 1980. The report reads, in part, "Was it a bear, or was it the legendary Big Foot or Sasquatch? Was it the Kooshtakaa?...Whatever it was, it was seen from the air by Stan

Figure 28: Photograph of tracks on a mountain in a *Wrangell Sentinel* article, Feb. 20, 1980, dealing with sighting of biped mountaintop tracks from air by two men in early 1980. *Photo:* Wrangell Sentinel

Hewitt and Chuck Trayler on a Wrangell-Ketchikan flight Monday, Feb. 11. The two pilots saw it atop a 2,900-ft. ridge, about 15 miles south of town, the last hill from town toward Thoms Lake.... As they circled for a better look, Hewitt said they 'saw something go off the edge' of the ridge. It was a big, black, grayish brown blob,' he said, but they couldn't spot it again after it went off the ledge." Just what was seen by whom was not made clear.

In a 2000 interview, Chuck Trayler recalled how they had flown low over the tracks repeatedly: "We had been on our way to Ketchikan on a sunny February day in late winter, early 1980, flying over a high ridge that crossed our flight path to Ketchikan, well away from any roads. There was lots of snow and the top of the ridge was well covered and corniced up to a steeply sloping snow face, maybe seventy-degrees steep. As we flew over, we were surprised to see a line of big tracks, apparently two-legged, coming out of the stunted trees and wandering around on the top. Circling repeatedly at 100 feet, we assessed the tracks to be evenly spaced, about a yard apart, and twelve to fourteen inches long, left and right.

"There were no reports of bears up yet, not even down low. We could see the beginning by the trees and what was most surprising, the end where they appeared to just end a few yards back from the edge of the cornice. Now, that cornice was not unevenly broken off anywhere and whatever it was, bear or sasquatch or man would have had a fall of perhaps up to a hundred feet before contacting the steeply sloping snow slope below. At that point the body would just be falling parallel to and in contact with the slope, not bottoming out for another four hundred feet down it seemed. And there were no visible marks on the slope where it should have slid down, and no body to be seen. You'd have to be pretty tough to make that kind of step-off, no matter whether you were a sasquatch or a bear. We circled at least twenty times and still couldn't figure it out."

Mr. Hewitt confirmed the pilot's recollection in a separate interview adding, "At that time we were looking at this from the point of view that excluding men and bears, theoretically these tracks could have been made by a sasquatch. We still don't know what made them."

Subsequent plane flights over the scene by Wrangell media, a biologist and others confirmed the tracks, and bear theories were put forward. Apparently no one managed to get in by helicopter or on foot for a closer examination. If it had been a brown bear, it is puzzling that a week later there were still no tracks returning to its supposed den site near timberline. [T80A-map3]

There are only two rather unlikely explanations for this report as given: first, the maker of the tracks jumped clear and survived the fall by landing a good ways down with no incident angle and sliding or rolling to a stop; second, the maker carefully retraced each step and returned the way he had come. It has been noted that bears will step in precisely their own tracks, man is known to do so and some reports of sasquatch tracks suggest they jump from dry patch to dry patch to avoid leaving more tracks than necessary. Some people have stated sasquatches are quite adept at hiding the direction of their travel. British Columbia guide and alleged sasquatch witness Clayton Mack described one set of sasquatch tracks that he had found in the Bella Coola region of B.C. as walking sideways on a snow-covered log for simply that deceptive purpose, then making a jump to shake off any pursuer. There is probably no satisfactory explanation to all this, but a sasquatch seems as likely a candidate as anything else.

Arnie, aged forty-eight, of Ketchikan is a talented Native artist and a big, genial man. In a 2001 interview, Arnie (not his real name) told me of three impressive tracks he had seen while deer hunting at a muskeg just behind Gem Cove, Revilla Island, on the east side of George Inlet, in August, 1979. Gem Cove is on the next peninsula southeast from that of Ketchikan, Saxman and Mountain Point. It is an unpopulated, unlogged and scenic arm of the island, favored by deer hunters but not many others. That day, Arnie had hunted up through hemlock, cedar and pine to the edge of a small muskeg that showed some tracks and sign.

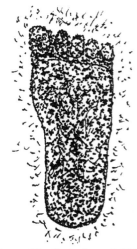

Figure 29: Illustration of 16-inch tracks as reported in muskeg near Gem Cove, Revilla' Island, AK, by a Ketchikan hunter in 1979. Sketch by author was under direction of witness. [T79B-map5]

"In soft, brown soil at the edge of the muskeg were numerous tracks, about sixteen inches long by six inches wide, wide at the heel and human in form, definitely not a bear's. And the first toe was the largest, no claw marks at all, and they would have shown. The tracks appeared to have come out of the forest about ten feet, stopped at the muskeg, and turned around, angling back into the forest. There were no bear tracks, no deer tracks, in fact there wasn't a darn sign of game anywhere there that August. There wasn't even a squirrel to be seen. But those big tracks were there. And they were an inch and a half deep. I jumped up and landed on one foot right beside one track and mine only sank in about one-quarter of an inch. In 1979 I weighed 315 pounds. What with no sign of game, and then seeing the size of those human-shaped tracks, I just gave up on the spot. The strange thing is that, the next fall, my brother-in-law got a deer in the same exact little muskeg. Besides deer tracks, he hadn't seen anything else."

Following the interview, Arnie drew an illustration of the tracks he had seen that August. His memory was kept sharp by his artist's eye and the incredible size and depth of the three clear tracks that had dwarfed his. [T79A-map5]

Figure 30: Illustration of tracks in snow, as reported by Klawock logger, on Flicker Creek north Prince of Wales Island, AK, 1985. [T85A-map2]

In an interview in 2000, Ted J., a forty-four-year-old Klawock logger and tugboat man, reported tracks two of his friends had found near the north end of Prince of Wales Island, Alaska.

"Around 1985, I was up finishing some shut-down work at a logging operation on Flicker Creek, halfway between Point Baker and Red Bay. Flicker Creek is situated two and a half miles east of Labouchere Bay. It was November and we had just shut down for the winter. Two friends I had logged with were staying on up there hunting deer, Bill N. and Doug V. It had snowed and I remember one morning Bill came tearing down to the camp and grabbed Doug saying, 'Come up the road [F.S. road 2087] and have a look at these big tracks in the snow!' These guys could look at bear tracks seven days a week and never blink an eye. There were a bunch of tracks, all about fourteen to sixteen inches long. The snow was only about an inch and a half deep and they showed up well. I believe Doug took a photo of the tracks with him back down to California."

On follow-up, nothing more has been found regarding the two loggers or any such photograph to date. [85A-map2]

Tracks in snow surfaced again about the same year, only this time about 100 miles to the east. Mr. Harvey Gross of Wrangell reported in 2000 that, in January, in or about 1985, he and his wife had seen peculiar tracks in the Wrangell area. They had been hiking in near Patrick Lake and were the only people there, three or four feet of fresh snow having fallen. The road they were hiking had been ploughed, but already several feet of new snow had fallen and accumulated. Mr. Gross said they had come upon, "fresh foot-long tracks, three to four feet apart in the snow, made by something walking upon two legs." He added, "It seemed peculiar because there were no other people in the area at the time." Although the tracks reported here are not large in comparison with usually reported sasquatch tracks, the unusual timing and circumstances of their deposition argue against a human origin. It seems just possible that Mr. and Mrs. Gross actually came upon the freshly made tracks of a juvenile sasquatch passing by the previous night. [T85B-map3]

Robert Durgan, a twenty-eight-year-old Ketchikan artist, formerly of Prince of Wales Island, recalled a set of large tracks which he, his father and uncle had seen on a deer hunting trip in 1986. On a snowy day in October or November of that year, the three hunters had been out less than a mile northwest of the village of Kasaan, on the east coast of Prince of Wales Island, when they were stopped dead. In the four inches of snow at their feet stretched a set of large two-legged tracks. "The tracks were about five-inches wide," Robert said, "and the two things we all agreed upon were that they weren't bear or human. The Native name all the old folks in Kasaan gave it was *gagiit*. We had all had bears try to steal deer we had shot, and were real familiar with bears. It was probably the only time all three of us agreed that the two-legged tracks we had seen were not made by any bear or a human. There's something undiscovered leaving tracks out there, that's certain." [T86A-map4]

Tracks in association with a reported sasquatch sighting are rare. One such track and brief sighting were reported in 1994. Mr. M. A., is a hunter and military man with whom I was acquainted in Ketchikan during the 1990s. When not at work or involved in family activities, he would spend much of his free time covering considerable distances over rugged terrain as part of a personal training program. In 1994, he told me of a single track he had seen in close conjunction with a brief sasquatch sighting above Carlanna Lake near Ketchikan in June, 1993.

"In June, 1993, I was getting in shape by doing some fast hiking with a light pack and over deer trails up the slopes around town. That day I had gone scouting up the ridge northwest of Carlanna Lake for deer and had made it up to a small saddle on the south side of the mountain at about 2,000 feet elevation. I was about level in altitude with some shoulders of peaks I had just gone up the week before on Gravina Island. This saddle was situated less than a mile northwest of Carlanna Lake up a series of deer trails. It was a clear afternoon.

"I had just stopped for a second, and took a second glance at what I had at first thought was a big, black stump about seven feet high. As soon as I turned to glance the second time I saw it quickly move on two legs as it disappeared quickly into the hemlock and second growth. Although I only saw it for a second, I saw it quite distinctly, as the distance was only about twenty meters [sixty feet] maximum. I kept moving along quickly along the deer trail and a few moments later, while still scanning for deer tracks, I saw a footprint. I had stepped right over it, stopped, realized what it was and took a good look. It was about fifteen inches long, six inches at the ball and about four, maybe four and a half, inches at the heel. The heel was definitely narrower. Most of the track was about three-quarters of an inch deep, parts were one to two inches deep. It was set in mud and clearly showed five toes. The toes were about two and a half to three inches long. It had a kind of hourglass shape, as though the heel and ball had imprinted more heavily. There were no claw marks. I didn't hear or smell anything unusual. On the way out in a creek nearby I also saw some six-inch trout. There were also some bait stations that hunters had set for bear but they looked old. I've been up Carlanna since but haven't seen anything more unusual." [S93B-map5](see Figure 31)

The next set of tracks to be reported gave the witnesses little doubt as to their owner. Mr. Bruce Shirley, a fifty-seven-year-old electrician from Klawock, Prince of Wales Island, recently reported finding two sets of large tracks near Labouchere Bay, at the northern extremity of Prince of Wales Island around 1994. I interviewed him in 2000.

"About six years ago, my son Scott and I had been deer-hunting at Labouchere Bay. It was the second week in November and pretty mild. There had been a skiff of snow between the rains, maybe an inch of snow on the roads. Up the Forest Service road just east of Lab' Bay, there's a road that runs south, and you can see the bay from there. We had gone up in a truck and were hunting up the mountain there. Just where the snow-line started on the mountain we found two sets of tracks, not of bears or humans.

"The largest set of tracks were about eighteen inches long, roughly human in shape but about eight inches wide at the ball. They were narrower at the heel—about five to six inches. You could see the five toes, the first toe was bigger, a bit longer and definitely wider than the others. Those toes were sometimes spread apart from each other and the snow showed up between them. They had a bit of an arch to them, or at least the snow was not pressed down so much there. They were definitely not a bear or a human. The second set was smaller, but I remember those were definitely bigger than a man's track. The toes would spread apart in some tracks. We followed those two sets of tracks right up the mountain and down over the other side, before turning back. We were the only ones up there. I'm real certain about those tracks, no doubt in my mind at all, they were sasquatch tracks." [T94B-map2]

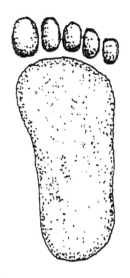

Figure 31: Illustration of a 15-inch track on a deer trail reported by hunter Mike A., west of Carlanna Lake, Revilla' Island, AK, in association with a brief sasquatch sighting report. Sketch by witness. [S93B-map5]

What is interesting about both the Labouchere Bay/Flicker Creek track reports was the previously noted sighting of a smaller sasquatch in the same area in August of 1987 by a young hunter, Mr. B. E. of Ketchikan.

The year 1997 again saw tracks reported in snow, less than forty miles south of Labouchere Bay. Mr. Jerry C., a Ketchikan hunter, recounted in an interview in 2000 that he and his hunting partner had seen a number of large biped tracks in fresh snow on northern Prince of Wales Island in 1997, which they felt were not those of a bear.

"The first week in November my hunting partner, T. A., and I had set up our camp four miles north of Naukati, Prince of Wales Island, where F. S. [Forestry Service] road 2058 turns south off of F. S. main road 20. There had been a fresh fall of new snow and we had road 2058, our usual spot for the last six years, all to ourselves. We had got up early and were about a quarter to a half mile up 2058 by 6:30 a.m. T. remarked to me as we arrived at the first muskeg there that straddled the road, 'What the heck are those tracks?' We had seen no tracks or sign and were curious. We got out of the truck and walked the two muskegs on either side of the road, each about the size of a football field.

"Leaving the timber on the east, crossing the muskegs and road

Figure 32: Author's illustration of an 18-inch track and smaller accompanying tracks in snow as reported by Bruce and Scott Shirley, south of Labouchere Bay, Prince of Wales Island, AK, 1994. [T94B-map2]

in about five inches of freshly fallen snow, was a line of huge two-legged tracks, each about fifteen and a half inches long by about six and a half inches wide at the ball, tapering back to about four and a half inches at the heel, with a stride of almost six feet. Most of them were about a yard apart. The straddle was about ten inches. There must have been about 150 of them. They were roughly the shape of a very large human foot, but showed no boot marks in the snow which had just stopped falling and was about five inches deep. The toes did not show but there was no arch to the feet like a human would have, just flat on the inside.

"They weren't bear tracks. Whatever it was had crossed the muskegs and headed into the timber on the west side. I remember T. and I were following them into the timber a ways when we both stopped and looked at each other as if to say, 'What do we do when we find what's making these?' and turned around and headed back.

"The year before at the same place we kept our hunting trailer down the road, I recall feeling the trailer shake one night and we both woke but that was all and went back to sleep. The next day one of four deer that we had roped hanging to the top edge of the trailer eight to twelve feet from the ground was missing. There had been no snow then and so no tracks, but we had assumed it must have been a bear. The rope had simply been broken by a pull or tug. I don't know about what took the deer for sure, but the tracks the next year in the snow were definitely not those of a bear."

Land management worker Eric E. of Ketchikan, detailed the following account of a single track found in fall, 1998 on southeast Revilla Island some ten miles northwest of Ketchikan in an interview given later that year.

"Chris M. and I were scouting for deer at the north end of the

Whipple Creek drainage, up a spur through clearcut that ended on a timbered knob at the 1,000-foot level. There was a little muskeg up there and we were looking at some deer tracks beside the muskeg. In a patch of thick brown mud beside the little muskeg we found a fifteen-inch track of a right foot. The track was very clear and showed five toes in a slight curve, less curve across the front than that a human foot, and no claws. The big toe was much larger. The track was several inches deep, much deeper under the ball. It was wide at the heel, not like a bear. It was similar to a bare human footprint but much larger, heavier and wider, especially at the back and middle. The print was facing north. The vegetation surrounding the track was typical muskeg, hemlock and some cedar. While hiking, we also noticed several bears and plenty of deer trails."

Halfway between Labouchere Bay and Wrangell lies the southern tip of Kupreanof Island. This was the location of a series of tracks in a second-hand report furnished by a Ketchikan contractor, Mr. Chris M. (not the man in previous account) given in 1999.

"This summer I traveled by boat through the rocks in Reid Bay, on the southeast of Kuiu Island, and interviewed its sole inhabitant, a trapper by the name of Rusty. The trapper told me that the previous July he had been staying alone in a float house at Little Totem Bay, on the south tip of Kupreanof Island, when he had noticed barefooted tracks, 'almost human-looking,' appearing overnight in the fine sand at the head of Totem Bay. He said they were about the length of a size-ten foot, but different." [T98A-map2]

It should be noted that, in Alaska, humans do not normally go barefoot anywhere outdoors, even in summer, except on a few of the more southern, warmer Alaskan beaches. Of course there are exceptions of the recreational variety, but not deposited overnight. In any event, the sound of a boat motor or floatplane would have announced the arrival of humans.

As in the previous report, the next set of tracks appeared in July on a sandy beach, but this time it was 100 air miles south. Frank S., Ketchikan, and his girlfriend Sara G., reported in 2000 of measuring an interesting set of large biped tracks on a sandy beach on the Alaska mainland, twenty-eight air miles southeast of Ketchikan in July of 1999. Separate interviews were agreeable in all respects. Following is Frank's account.

"We had taken my boat on a sunny day in early July to a beach

Figure 33: Illustration of one of 150 tracks in snow. The tracks measured 15 ½ inches and were reported north of Naukati, Prince of Wales Island, AK, in 1997, by Ketchikan hunters J. C. and T. A. Sketch by author under direction of witness J. C.

on East Behm Canal about halfway between Point Sykes to the north and Boca de Quadra Inlet to the south. Anchoring very close to a thin mainland peninsula in a sheltered spot about two miles south of Black Island, we went ashore in the skiff.

"As soon as we pulled up on the sandy beach I spotted this string of large tracks paralleling the water. It was the first sunny day in a week and they were fresh. I was surprised to see that they were roughly human in shape, but six inches wide at the ball by fifteen and a half inches long. My girlfriend watched as I tried to jump from track to track. They were very close to six feet apart and I had to jump as hard as I could to get from track to track. They were four inches wide at the heel and had no arch. All five toes showed clearly and were approximately two inches long. The first toe was barely longer than the second but about a half-inch wider. The toes weren't separated, the white sand showed clearly that the ball and toes swept back to the outside in a shallow arcing line. They were about an inch to an inch and a half deep. My tracks were about a quarter- to a half-inch deep.

"The tracks followed the shoreline south, before turning at a right angle and going up into the trees. The total length of the trail of tracks was about seventy feet. There were no other boats in the area and the beaches along there were all beautiful and secluded, great spots for clamming or beachcombing. We looked at the tracks for a while, then rowed back to the boat and continued cruising for the rest of the day." [T99A-map3]

There is sometimes mention made of tracks seen over a period of time in one locale. This information is usually shared with close friends or members of a family, but not reported or documented, such as in the case of Mr. S. B. of Ketchikan, who stated the following in an interview in 1999.

"From 1994 to '97 our family and relatives have regularly gone hunting bear up the timbered slopes above Yes Bay, at the north end of West Behm Canal. While there, we have often seen fair numbers of large humanlike tracks. We have also heard a lot of screams, which we don't allow to be any commonly known type of animal. We don't usually talk about this to anyone." [T94A-map3]

On at least one occasion, tracks have been found that seem to walk directly into the ocean. Commercial fishermen Jeff J. and Steve S. reported such unusual tracks by the Portland Canal along the U.S./Canada border in January, 2000. Steve recounted the following.

"On January seventh, we were just fifty feet from shore on the American side of the Portland Canal north of Pearse Island, B.C. This was at a spot just north of Reef Island, which is just a few hundred yards off the Alaska mainland. On the steep, snowy beach, where the forest comes down real quick to the water, there was a line of big, two-footed tracks coming straight down out of the trees to the shoreline. There were maybe twenty tracks. It was high tide. There was about four feet of snow and the tracks showed no drag marks, just punched in, left, right, left. The country is too steep and thickly forested for moose, even if the tracks weren't two-footed. I have a commercial guide's license and have a pretty fair idea of what they weren't. Definitely didn't look like a bear. No belly marks in the deep snow. It must have had long legs. The tracks also just went straight in, didn't parallel the shore like a bear or moose would. There were no tracks coming out there. I don't know where they came out. It just looked like something big on two legs wanted to go in for some abalone." [T00A-map6]

There is no proving that these tracks were those of a sasquatch, but Steve, a commercial guide, quickly discounted bear and moose. Since there is no human habitation for miles along the canal, there is just as good a possibility that these tracks were made by a sasquatch foraging for seafood or making the short transit across the narrow canal to Canada.

Tracks of a larger size were more recently reported closer to Ketchikan in July, 2002. Mike V., girlfriend Emily G. and a friend, all from Ketchikan, were beachcombing a small forested island at the mouth of White River, a half-hour boat trip up the west side of

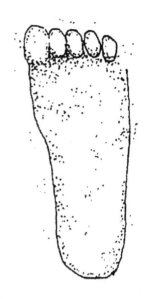

Figure 34: Illustration of a single 15-inch track in mud in Whipple Creek drainage, southeast Revilla Island, AK, as reported and sketched by land manager E. E. in 1998.

George Inlet, northeast of Ketchikan. Emily reported they found huge fresh tracks all along the sandy shore of the island. In a later interview she commented:

"Mike called us over to look at these huge tracks. The size of the larger set was about 20 inches and the smaller was about 16 inches, with the larger set having a distance of about five feet apart. They both showed five toes, with the first toe the biggest, and they were made by something a lot heavier than us. The tracks were all two-legged. Both sets of tracks led along the shore of the island...seemed to go all around it. There was one other interesting thing," added Emily. "At that time there were a lot of harbor seals with pups on the rocks around the small island. It made us wonder if that had anything to do with the creatures making those tracks being there." Besides seals, the usual healthy numbers of deer and salmon were reported in the area during the summer. [T02D-map5]

Recently, there has been an interesting discovery in the field of dermatoglyphics (fingerprinting science), that has a bearing on sasquatch tracks and their casts. Anatomist and primate locomotion specialist Dr. Jeff Meldrum of Idaho, and police fingerprint specialist James Chilcutt, Texas, have demonstrated an unusual dermal ridge pattern in sample track casts, put forward by various investigators as possible sasquatch tracks. Generally summarizing some differences, Professor Meldrum noted in September, 2000: "Apes tend to have ridges that diverge from a delta on the lateral side of the foot and a significant proportion of them run longitudinally, whereas in most humans, the ridges run transversely. We observed repeated instances of ridges running longitudinally, parallel to the borders of the sole, whereas in humans, they are generally perpendicular." This apelike characteristic of detailed sasquatch tracks

Illustration: Patrick Beaton

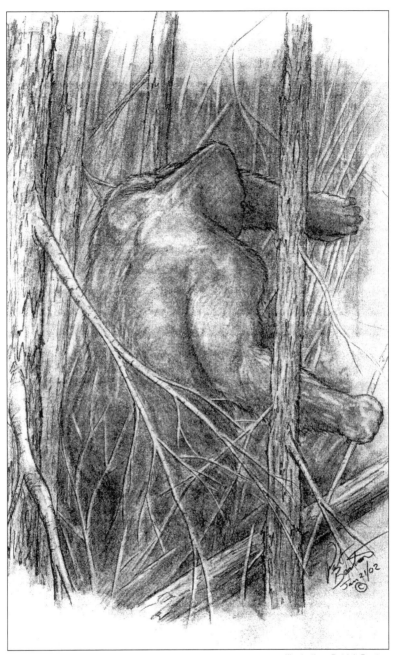

Illustration: Patrick Beaton

Illustration: Patrick Beaton

Illustration: Patrick Beaton

becomes one more item of evidence that, while not proving to everyone that sasquatches are real, places them more on the ape side of things, a subject to be dealt with in its own chapter.

There is one confidential sasquatch sighting report by a family driving over Whipple Creek Bridge at 5:00 p.m., March 1, 1998. Father and step-son agreed the dark brown hair-covered creature was about seven feet tall, and walked north to the highway center line about seventy yards in front of them, then ran the thirty remaining feet to the forest in just three bounds. The man's five-year old daughter also said she noticed the creature from the backseat and had called out "Bigfoot!" The stepson added that the hair on the creature's "backside" (sic) looked "real matted and foul." The forest on both sides is old mixed second-growth. The family is known to me personally and the father, a professional man who has a background in the biological sciences, wanted to stop and look for tracks, but the step-son was in a hurry to get home and the man was not able to return until the next day at noon. As the rain had not let up, there were no tracks or sign to be found in the ditch or elsewhere. [S98A-map5]

On February 3, 2002, a set of tracks was reported at a location two and a half miles northeast of Whipple Creek. In an interview given in April, Ketchikan delivery driver Andy Rasmussen reported that he had seen a string of unusual tracks off North Point Higgins Road, a heavily forested area fourteen miles northeast of Ketchikan. Mr. Rasmussen, a thirty-year-old outdoorsman born in Ketchikan, stated:

"Early in the afternoon of February 3, I was making a delivery on North Pt. Higgins Road, just past Mattle Road, and the homeowner and I had to go to the back of the house through snow. I noticed an unusual chain of two legged tracks, about three feet apart, crossing in front of us and heading forty feet straight into the trees. We had to step through them and I was looking at them the whole time, not saying anything to the man behind me. The tracks were as deep as the snow, about six inches, and were longer and wider than my size eight boot, even at the heel. They were almost as wide at the heel as at the front and weren't made by any bear on two legs or human boot, I'm 100 percent sure. What really got my attention was the impression in the snow around each track of matted hair or fur. That was very plain. Although no clear toe marks were visible, the snow beneath each track was heavily compacted and flat

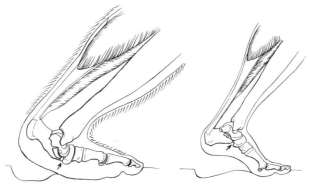

Sasquatch versus human. Illustration: Dr. Grover Krantz

all along, no lines across at the heel. They looked fresh, as though they had been made the night before.

"I was puzzled the homeowner hadn't said anything. He had moved up to Alaska only a year ago, and was just walking along behind me without saying a word. I left it at that. I only know they weren't made by a human or a bear. I hunt and that, and I've lived here all my life, but I've never seen anything like those tracks in my life before." [T02B-map5]

Three nearby highway sightings, two recent, lend support to the recent track reports. Fifteen-inch tracks had been reported in the Whipple Creek drainage twice that same year, once by an off-duty land manager, and once by a teenager who reported what he thought was a huge bear track on a logging road north of a security trailer. Loggers working above Whipple Creek in the summer of 2001 had remarked how, on returning to their work site a number of mornings, they had noticed small items of equipment rearranged, and some things missing. During military exercises at Whipple Creek in March, 1998, several local national guard members commented that they had been surprised by an angry marine sergeant who allegedly accused them of dressing some large local up in a gorilla

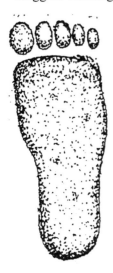

Figure 35: Illustration of one of a series of 15 ½-inch tracks in sand reported north of Boca de Quadra Inlet on the mainland side of East Behm Canal, AK, by two boaters exploring beach in 1999. Sketch by author under direction of witness.

suit and running at them across the creek. The local men had found this amusing to say the least, whatever truth there might be behind such a tale.

The highway in the area of Whipple Creek has been the location of at least three documented sasquatch sightings, dating from the late seventies [S70B] to 1998 [S98A], [S98B]. While there is nothing conclusive to link any of the reports, it is interesting to speculate that Whipple Creek is home to at least one sasquatch, and that neither logging nor military exercises have had any effect in changing that.

In early 2002, Mrs. Jodi Muzzana, Ketchikan, commented on strange tracks at Herring Cove, Revilla Island on the road southeast of Ketchikan, the location of several sightings the previous year. In an interview she stated she had been puzzled by one of the tracks, just over fifteen inches long and quite broad, showing in the wet silt between the highway guard-rail and the edge of Herring Cove. She had been taking a morning walk with her husband, early March she thought, just as the last snowfall was melting and had stopped to look at a cluster of large tracks in several inches of snow on the bank that slopes some thirty feet down to the water beside the bridge. Jodi said, "What was peculiar was that these tracks were barefoot, much larger than a person's, and showed the five toes. They weren't a bear's tracks, we have lots of them around, and I called my husband to come and look at the clearest one. He wasn't real pleased with the whole thing, it was surprising how close to houses the tracks were, and all. I saw some dark hair by the track and picked up a strand to look at along the highway, but it got blown away. We just continued our walk and didn't concern ourselves much more about it." [T02C-map5]

While one initially might expect the ratio of sasquatch tracks to sasquatch sightings in southeast Alaska to be higher than reported (more than fourteen reports mentioning tracks, approximately sixty sighting reports) it must be remembered that much of southeast Alaska is covered with very thin forest soil that does not register tracks well. Blacktail deer and bear often use trails of their own making, but sasquatch tracks seem to be found only rarely on such terrain. This may be because sasquatches do not forage in such small areas as a blacktail deer (usually one deer to four square miles) and are perhaps more able to simply barge through wherever they wish. Or perhaps, as some reports suggest, they are intelligent enough to usually avoid such obvious surfaces as trails and may even habitually step only on surfaces that conceal their passage or direction of travel. At any rate, the tides and rain will have their way with most tracks of any type along the coast, and considering the population of black bear—two to three thousand on Revilla' Island alone according to Alaska Department of Fish and Game—it is amazing how few bear tracks are found away from garbage sites each year. It is also astounding how quickly 140 inches of annual precipitation will "wipe the slate clean" of tracks, so to speak. At any rate, reports of tracks matching the description of sasquatches' are regularly seen in southeast Alaska, and one may be certain that there are many more such reports that are never disclosed.

Alaskans themselves are often as retiring as the subject of this book. Although an intrepid few venture far off the logging roads, coastline or the shorelines of "flyable" lakes, most spend their summer or fall weekends only as far as their vehicles or boats can get them. Considering Alaska's rugged terrain, this does not cover much of the country, and since logging has been scaled back, it is not likely to get much greater. Since few Alaskans are actually on the beaches, especially in winter, it is not surprising that more tracks have not been reported. It is hoped, that with increasing awareness of sasquatches as an animal that lives, breathes and leaves marks of its passage just as humans and bears do, more will come to light. Until then, it may be that the occasional "bear" tracks found on winter beaches will, on second glance, prove to be quite a bit more than what they may have at first seemed.

1. Krantz, 1999, *Bigfoot Sasquatch Evidence*, Hancock House Publishers, Surrey, B.C., pp. 52–86.

2. Meldrum, Jeff, personal communication.

3. Sheldon, Ian and Tamara Hartson, 1999, *Animal Tracks of Alaska,* Lone Pine Publishing, Renton, Washington and Edmonton, Alberta, pp. 96–97.

4. Simons, E., *Gigantopithecus, Scientific American*, 1968, June, pp. 40–46.

12 Screams in the Forest

"I heard a terrible yell, louder than a person could make, from the forest, and louder than a bear could possibly make. I have trapped and run into brown bears all my life. The slopes around the cabin were logged for 500 yards around and it had to be coming from beyond that. The first yell was about fifteen seconds long and it repeated itself two more times over the next minute. It was like a scream, a yell and a roar all in one. To me it sounded like it was in a lot of pain. I know it was not a bear in pain. Those yells had quite a deep pitch to them as well. In all my years up there, I never heard anything even close to that. I took my rifle and went out there and looked around but couldn't see anything, even tracks!" Mr. John Kristovich, Portland Canal, Alaska, 1997

Reported vocalizations possibly made by or attributable to sasquatches have been described by researchers Bindernagel[1] and Green[2]. Green also provided an analysis of types of reports of sasquatch vocalizations from various locales in North America, with earlier data, in 1980. Following is Green's summation of the relative frequencies of groups of vocalizations.

"The only sounds that are consistently reported everywhere are screams, 37 all told, and whistles, only 10 reports but well distributed. Grouping somewhat similar sounds together, there are 50 screams, yells and howls; 23 whistles, squeals, blowing sounds and wailing cries; 17 grunts, growls and coughs, eight chatters, barks, mutters, laughs and yelps. Some of the animals that screamed did so after being shot, but it is still the most common sound, even without provocation, in every area."[3]

Of vocalizations reported in southeast Alaska that might be attributed to sasquatches, thirty-five are documented in this book, mostly in this chapter. Most often reported are screams or roars of extremely loud volume (16), yells (9), and occasionally crying noises (2), chattering (4) growls (1) and grunts (1). Loud whooping has been reported reported twice. These audible vocalizations and noisemaking, unlike those made by known species with which outdoor Alaskans are very familiar, are often of incredible volume. These reported vocalizations are documented here in a chronological order.

Some of the earliest references to vocalizations that might be attributed to sasquatches appear in accounts published by the late Alaska pioneer, A. H. ("Handlogger") Jackson, whose book *Handloggers*[4] details an interesting series of encounters with unknown calls heard over a period of time. The location of these events was a very narrow isthmus of land on the middle of the northern shore of Revilla' Island and its peninsula, Claude Point. The following is excerpted from Jackson's book.

"Back in 1921 Jack McKay was operating the boat Salmon Bay as a camp tender for the Sawyer McKay logging camp in Saks Cove. Once when he was alone on the boat on his way to Ketchikan, evening overtook him at Claude Point so he went into the north Harbor to anchor for the night.

"Jack had just dropped anchor, when there came a call from the outer end of the spit. While he was making the anchor line secure, he answered. Just as he straightened up, he heard a second call.

"'All right,' he called back, and started aft. There came a third call. He looked the beach over carefully, saw no one but replied, 'I'll be right in.' Still no one appeared on shore.

"Jack lowered his skiff and rowed ashore. There was no one in sight. He shouted several times but got no answer. Thinking someone anchored on the other side of the spit had walked across, he landed and walked far enough that he could see the other harbor. There was no boat and no one ashore. He looked around for a camp but found none, nor did he find a human footprint, nor any sign that anyone had been there for months.

"Back aboard, he watched the beach until dark and again in the morning, but saw no one and heard nothing more. On his subsequent weekly trips he watched the shore closely as he went by, but never did see a sign of a human being.

"Jack McKay was an old-time woodsman, not easily confused or mystified. [VO21-map5]

"About ten years later [around 1931] Jim Pitcher, who had not yet heard Jack McKay's story, had almost an identical experience. Cruising alone on his boat, he anchored for the night on the opposite side of the spit, heard the same three calls, rowed ashore, and found no one. He too thought someone was anchored in the other harbor and walked across, to find it deserted. Again there was neither camp nor footprint.

"When Jim told me about the experience he said there had been no words to the call, it was 'just a shout, like any Swede fisherman calling to attract your attention.' He declared that in more than fifty years in the wilderness, this was the only thing he had ever encountered that he could not explain." [VO31-map5]

Jackson then goes on to recount events in the summer of 1944, when he and his wife Ruth were at Claude Point. At that point they had known nothing of McKay's or Pitcher's experiences. Jackson details how one morning, after they had been staying at Claude Point logging and fishing for a month, Ruth had got her tackle and rowed out into the bay.

"Just as she stopped to drop her lure into the water, she heard a call from the shore across the bay. That was nothing unusual, for often when I was cruising out more trees to cut, I would come out through the woods. If I was closer to her than my skiff, I would call out for her to get me.

"Ruth had just turned her skiff to look for me when she heard the call again, directly in front of her. It didn't come from the beach. It came from the steep mountainside about 200 yards back in the timber. Her heart leaped into her throat as she pictured me lying up there in the woods, hurt and needing her help. She answered that she was coming. Again the call, loud and clear. She noted the exact spot and leaned on her oars.

"Abruptly she stopped, dumbfounded. From somewhere down toward Beaver Creek came the steady chopping sounds of my axe. In dismay she looked from one spot to the other. She could not see into the timber, but anyone up there could see her. She called repeatedly. No answer."

Jackson then concludes by stating that his wife then checked the other side of the spit as McKay and Pitcher had done and after returning to the launch had waited for him, who was as mystified as she had been. They checked up on the timbered slope, thoroughly searching every game trail and around every tree, but found no sign. Nor did they encounter anything similar the whole summer they camped and hunted there. In Jackson's words, "The voice of Claude Point remained a mystery." [VO44-map5]

It may be of some interest that on each of these incidents, separated by decades, the calls were each reported as heard in groups of three. No owls, especially the diurnal ones inhabiting Alaska, are known to call in such a fashion, or as loud and with such a human timbre to them. What is especially interesting is that some of the loud calls or rapping noises mentioned elsewhere as possibly those of a sasquatch have also been heard in a series of three.

Others have reported hearing loud humanlike yells at Claude Point. The late Bruce Johnstone Sr. stated to me in an interview in 1998, that he had been camping very much alone, at Claude Point around 1944, and remembers hearing a single loud humanlike call. "It didn't sound like it was hurt," he said, "it sounded like it was trying to communicate."

Mr. Johnstone had spent probably as much time in the bush of southeast Alaska as any man still alive, and was well familiar with all the vocalizations the local fauna can make. He cited two more occasions where he noted peculiar calls in the bush. One was while camping near the mouth of Bakewell Creek in the Smeaton Bay, Misty Fjords area near the southern tip of the Alaskan panhandle, in summertime around 1948.

He had just let his fire go down and thought it was perhaps midnight, when he heard noises in the bush behind him, near sea level. "It wasn't like porcupines" he said, "there seemed to be two of them, and they were talking back and forth. It wasn't quite human. It almost sounded like, well, have you ever heard two old Finlanders chattering away?" In the morning he found no sign of porcupine, and the altitude was much too low for ptarmigan. He stated that he found one peculiar-looking footprint unlike anything he had ever seen, but couldn't make much of it. Lest one imagine for a moment that Bruce was unfamiliar with bears and their tracks, either brown or black, one should note that he has been well documented for a lifetime's involvement with bears. He not only survived a charge by a brown bear, but had taken good care of himself with them at close quarters on a number of occasions. [V48A-map3]

In 1999 Johnstone told me that the most interesting account of unusual humanlike calls in the Alaska bush is one he experienced while deer hunting in the muskegs above a bay on the south side of Chomondoley Sound, Prince of Wales Island. This bay has often been referred to by locals as Wild Man Cove. Mr. Johnstone stated that in the fall of 1952 or thereabouts he was staying on a boat with Virgil Crosby, who was a federal fish and wildlife officer, A. H. Jackson and others. Deciding it was a good day for venison, he headed up several miles through muskeg-dotted timber.

"I hiked straight back up about a mile or so to some muskegs for deer. They were really lovely deer hunting spots. I came to this one muskeg and close by I heard a call; it sounded like a loud 'Heyyy!' I called back, but no reply.

"That thing moved all around me, calling, circling but never coming out. This went on for over twenty minutes and by this time I was thinking, 'All I have is my deer rifle, and I've put up with this all I'm going to...' I got out of there, I quit and went right back on down to the boat." [VO52-map4]

Following are other vocalizations that could quite probably be categorized as Sasquatch vocalizations or noisemaking heard in southeast Alaska. They have been recorded chronologically.

• 1948 (approximate): A second-hand story involving possible sasquatch vocalizations was passed on to me in 1997 by Mrs. Barb B., Ketchikan, who knew the informant well. "He was an old-time local hunter and commercial fisherman who was born and raised in Ketchikan. He appeared for coffee one summer around 1947, and in telling this story," she noted, "you could tell he was upset." Barb recalled his account in his words.

"I was just out hunting at Carroll Inlet [Revilla' Island], right at the tip of California Head. I was tied up right at shore and it was getting dark, when all of a sudden I smelled a real rank odor, whereas bears have a real musky odor. I've been hunting here close to forty-five years or so and I should know the difference. Right then I heard rustling noises, but it was real all right. I don't believe in any of the old *kushtakaa* folklore! Then I started hearing a real strange grunting noise in the bush about ten yards from where I was tied up at. It was no bear! And that's when I got out of there. It was almost like I didn't want to know!" [V48B-map5]

It is of course possible that grunting could have emanated from a bear, but on the other hand, one ought to lend some credence to a longtime hunter who, in all probability, knew well what a bear sounded like, and who asserted it was definitely not that which he had heard.

• 1950: Researcher Al Jackson, of Klawock, reported in an interview in 1997, "while commercial fishing in the summer of 1950 with my father on his boat near the head of Soda Bay on the west coast of Prince of Wales Island between Hydaburg and Craig, we both heard repeated loud screams from shore nearby. 'That's not a bear,' my dad added." [VO50-map4] Jackson himself has little doubt about what he heard. "It was just too darn loud to be a bear," Jackson commented. "I'm pretty certain that what we heard had to have been a sasquatch."

At least four documented West Coast reports make mention of chattering. In the case of the classic 1928 Albert Ostman report from Toba Inlet, B.C., the sasquatches were apparently communicating back and forth with each other while making the sound. The following report is a detailed account of just such sounds, and the ensuing "conversation" that three men are reported to have had with an unseen creature. The event was located on the Unuk, a large glacial river on the mainland north of Revilla' Island, flowing south out of the mountains in British Columbia.

• 1956: "On the thirteenth of September in 1955," recounted Bruce Johnstone Sr., in a 2002 interview, "two days before the opening day of moose season, I was on the Unuk River with two partners, Ed Roberts and 'Tongass Pete' Bringsli. We were camped at the bottom of Lake Creek on a little island called Bishop's Island. At the time, over forty-five years ago, there was only one small patch of trees on the northwest corner of the island, not all grown in like it is today. We had our camp there and had a good view of the timbered banks all round us and were quite comfortable.

"Just as it was getting dark, we heard a noise coming from the far bank, it sounded like a rising and falling series of barking, chattering sounds. We answered back, but it waited a minute before answering and was moving along the edge of the trees. It was wailing and making different sounds, and I asked Ed, who had a lot of experience down south with coyotes, if it was a coyote but he said not. And we kept calling back and forth with it every minute or so for the better part of an hour.

"The sounds were all jumbled together and it sounded like whatever it was, was trying to put words of sorts together, like it was trying to communicate with us. This would go on every minute or so.

"Whatever it was circled around our camp in the forest without ever coming out. It sounded like it was trying to talk to us but didn't quite have the nerve to step out and let us see it. It wasn't real high pitched and was about as loud as we were, like a man talking in a normal voice.

"The individual sounds themselves sounded a bit like the sounds made by Tahltan Indians I used to hear long ago, but it wasn't any of the Native or white languages I have ever heard. None of us could figure it out and it quit just as it got dark. It wasn't a porcupine or any bird I have ever heard in these parts, it was a bit like the chattering I heard in the bush one night below Bakewell Creek on Smeaton Bay, Misty Fjords, but a bit lower pitched. We just didn't know what it was at all. I've never heard anything like it since." [VO56-map3]

Mr. Johnstone's two reports mentioning a speechlike chattering should be noted for further comparison with the 1977 acoustic analysis of the Berry-Johnson High Sierra tapes, which will be described later in this chapter. These are all suggestive of a more complex system of vocalization, and are dealt with later in this chapter.

• 1958–60: Mr. A. K. of Ketchikan and formerly of Klawock, noted to me in 1999 that on several occasions in the late fifties around Klawock, Prince of Wales Island, he and others "had heard loud yells and screams in the bush there that no one experienced with hunting would attribute to bears." He stated that he and others did also not feel they were of human origin. [VO58-map4]

• 1968: While not a report of purported sasquatch vocalization *per se*, the following account aroused my interest. Portland Canal commercial fisherman Patty Kristovich Jr., recounted to me in 2000 that he had heard something, not a bear or moose, in the timbered flats three miles northwest of Hyder on the Salmon River in 1968 or thereabouts.

"I was out with my dad in November, we were hunting and there was no snow down low. There were no roads or people on that side. You had to get there by boat, and we were alone. It was early in the morning and we were just going along when we heard something louder than a bear or moose moving really close by. We all grew up with big bears and moose, you could tell this was different and besides all the bears had lit out higher up for denning. There was no bear sign anywhere all around down there. We just asked each other 'What the heck is that anyway?' and waited. Circling around later, there was no sign anywhere. Whatever it was could sure hide its tracks!

"On another occasion near there my dad shot a brown bear and it looked as if something had twisted its lower jaw almost off and smashed its front paw. There were no bite marks, claw marks or wounds at all as you would expect if it had been fighting another bear. It gave us a curious feeling."

• 1970 (approximate): Harvey Gross, Wrangell, reported in a telephone interview in 2000 that sometime around 1970 he was with relatives on the Stikine River near Wrangell. He stated: "We were camping near Cottonwood Slough, up the mountain along the river on the east side, about four miles from the river mouth. We all heard loud screams, louder than a cougar, repeated three times. They were about as loud as a chainsaw. Although three cougar have been shot in the area, we did not think it was a cougar scream—too loud." [V70A-map2]

• 1974: Al Jackson, Klawock, stated that on Labor Day, 1974, while deer hunting one evening with friends and family in old-growth timber near Swan Lake [above the head of Carroll Inlet, Revilla' Island] he heard a "loud screaming yell unlike a bear in pain," and the sound of heavy breathing from the thick undergrowth. He added, "There were no other groups up there at the time. I've heard all the other animals' calls around here and it was definitely not any of those."[V74A-map5]

• 1975: Throughout the 1970s, southeast Alaska residents continued to be puzzled and alarmed by strange vocalizations. Mrs. J. M., Ketchikan, said in an interview in 2000, that in the late fall of 1975 her late husband and her sons had been fixing up a float home at the head of Saltery Cove on the south side of Skowl Arm, east Prince of Wales Island. It was about 10:00 p.m., dark out, and the family all heard a series of loud screams that lasted ten or fifteen minutes. The sons all thought the screams sounded like they were coming from quite a distance away. Mrs. M. had heard screech owls and was sure that was not the source. [V75A-map4]

She added that her husband and one of the boys had also heard a scream from the forest behind the beach near there on one occasion while cutting up a beached cedar log for firewood. That scream had been from the woods close by, just behind where they were working. [V74A-map4] The spring of the previous year, 1974, her husband had been digging for a water source at a stream beside the muskeg just behind the float house. He told her that evening he had smelled an animal, and at the same time seen a single large track in light snow at the same spot. "He told me the hair on the back of his neck went right up, he had the sensation of being watched, then he got the hell right out of there."

• 1977: The late Harold Omacht, Ketchikan, stated that while hiking up to Granite Basin in thick forest six miles north of Ketchikan in summer, 1977, he heard a series of sounds that were quite loud. "They sounded like a kind of loud chuckling from the thick forest upstream of me, but much louder than what a ptarmigan might make. I was going to take landscape photos of Granite Basin, but it was quite unsettling. So after hearing the calls I turned back to town." [V77A-map5]

• 1978 or 1979: Rod Stockli, Ketchikan, stated in an interview in 1999 that he, Bill Olson and others had heard extremely loud calls from the dense timber across from their campsite at Last Chance Campground below Connell Lake, Revilla Island, one summer night. This exact location is also mentioned in two sighting reports, one around 1978 and the other in 2001.

"It was dark and the windows in the Buick were up and shut out all the outside sounds. Suddenly we heard a scream that was louder than any animal I've ever heard, coming from the slope just across the creek, maybe fifty yards away. It was long and loud, almost like an elephant trumpeting. It wasn't any animal I've ever heard in the bush and I was born in Alaska. We decided not to stay." [V78A-map5]

The 1980s witnessed four more reports of loud peculiar calls, three of them screaming, and one was a series of loud, whooping sounds.

- 1982 or 1983: Norm J., Ketchikan, a commercial fisherman and Native carver, recounted an unusual call he heard off the southern tip of Gravina Island in July while fishing for silver salmon.

"It was ten o'clock in the evening and I had anchored fifty yards offshore near the head of Dall Bay. There was a deserted cabin right there. A friend of mine had been staying there a while but was away at the time. The stars and moon were out and it was a warm night. As I was standing on deck, the calm was shattered by a bone-chilling scream. It was one long call. It started loud and stayed loud, louder than a human could yell. It seemed to be coming from the edge of the trees onshore. I was on the boat, I knew it wasn't going to hurt me. It couldn't have been an owl, a deer or a bear—too loud. It sounded like something dying. It was a scary sound!" [V82A-map5]

- 1986: An experienced Ketchikan hunter and guide (name withheld) told me in a 1999 interview that he and two longtime hunting companions, R. D. and M. P., had gone north into the mainland mountains above Revilla' Island for their annual Unuk River moose hunt.

"It was late September, '86, getting dark, and we were floating quietly back downstream above the Unuk River, about one quarter of the way up Lake Creek.

"Something had just spooked a large flock of geese from the nearby marsh to the northwest and suddenly we heard an extraordinarily loud whooping scream from just inside the thick alders. It was about two seconds long, about half as loud as a jackhammer and sounded like what I can only describe as a monkey whooping. There were three calls, each about thirty seconds apart. It was like no other animal I have ever heard in Alaska. If it weren't for the loudness, I would have said it sounded like the kind of call made by a chimpanzee." [V86A-map3]

However strange this might sound, a five-foot male chimpanzee has quite a distinctive roar, besides the usual whooping and hooting often associated with them. This apelike whooping has also been noted personally by wildlife biologist John Bindernagel near Strathcona Park on Vancouver Island in 1982, and in 1993 on Prince of Wales Island, AK, by a forestry professional in a report to follow.

• Circa 1986: Mr. M. M., of Ketchikan, whose extraordinary report of a female sasquatch in Ernest Sound is recounted in the previous chapter, told me the following in 1999.

"I was goat hunting in the fall of '86 at the end of Smeaton Bay in Misty Fjords National Monument Wilderness. I was up above the canyon east of where the road ends when I heard a real loud scream, not like a hurt bear; not a cougar screaming. It just wasn't like anything I've ever heard in the bush before. My feelings are that it had to have been a sasquatch." [V86B-map3]

• 1988: Leroy J., a Ketchikan hunter, stated in an interview in 1999 that he had heard strange vocalizations of great volume north of Wadleigh Island above Klawock, Prince of Wales Island.

"Around October, 1988, my friend D. L. and I had been hunting deer. We had anchored my boat north of Wadleigh Island in the Shinaku Inlet area and we had come around this cove to hunt. We had four deer. D. was around the corner and I was at the boat. We were tied up close to shore and had a real good situation, getting our deer onto the boat easily. It was getting dark when from out of the bush I heard these three long, loud screams. They almost were like a human scream 'Waaauuugh!' But three times louder! It wasn't an otter and it sure wasn't a bear. My friend came around the corner and he had an amazed look on his face. 'Did you hear that?' he asked. I could feel my hair standing up. He agreed it wasn't a bear. We left." [V88A-map4]

Over the past ten years, an increasing number of southeastern outdoor people have reported loud vocalizations, several times in association with "wood-on-wood" rapping sounds.

• 1992: As Bruce C., a Ketchikan hunter and commercial fisherman, attested in an interview in 1999: "In July, 1992, I was camping with my brother about one and one half-hour's drive up a logging spur road off the west side of the Hydaburg highway on the southwest part of Prince of Wales Island. The road goes west up towards Lake St. Nicholas. It was dusk and we were just starting to set up camp at the end of the road in thick timber.

"Just then we heard several real loud screams from nearby. Each was about ten seconds long. It sure didn't sound like a bear. We just looked at each other, got in the truck, left our tent and gear and left! We didn't drive back up until the next day when we collected our gear and got out of there." [V92A-map4]

Cougar have been reported to Alaska Fish and Game only once on Prince of Wales Island. While they do scream when mating, ten seconds is a long time for a cougar to wail. And sasquatch reports are much more common than cougar reports throughout southeast Alaska.

• 1993: Mr. D. H., of Craig on Prince of Wales Island and an experienced forestry professional, reported that while scouting the country above Old Frank's Watershed on central Prince of Wales Island with Mr. T. T., another forestry professional, they heard some very unusual calls. "It was daytime and we were near the edge of a muskeg on the height of land. From the trees, a hundred yards or so away, we heard a series of loud calls," which he described as "somewhat like a monkey." [V93A-map4]

• 1994: Rob Sanderson, Ketchikan, stated in 1994 that he was fall hunting above tree line above Eek Inlet, southeast of Hydaburg on Prince of Wales Island. It was early in the day and he was sitting at the edge of a high meadow blowing a deer call, when he heard four rising whistles, each three-seconds long, over a thirty-second period. Ten minutes later this was followed by a wooden rapping sound from the trees. He said, "When I heard that wooden rapping sound, I felt the hair go up on the back of my neck. It wasn't like anything I had ever heard in the bush before. After a bit I headed on down and didn't hear it again." [V94A-map4]

• 1996: Larry D., Ketchikan, reported to me that he and his uncle, both experienced Native hunters, had been camped at Bostwick Inlet on the south end of Gravina Island west of Ketchikan, fall deer hunting two miles up Bostwick Creek, when they were startled at hearing a loud scream nearby in thick vegetation.

"We were both familiar with otters, bears and the variety of sounds they can make. We agreed it was neither. We just both looked at each other and headed back to camp. We concluded there were a lot of other real good places we could hunt. When we eventually got back to camp at the beach, something had vandalized the camp, really messed it up."

This report, and ones like it, suggest a theoretical connection between reports of sasquatch screams and possible threat displays or territorial behavior.

- 1997: Tombstone Bay, halfway up the Portland Canal, on the American side was the site of several reported vocalizations purportedly made by a sasquatch. John Kristovich, a retired fisherman who has lived most of his life at Tombstone Bay, trapping and commercial fishing, told me in a telephone interview in February, 2000, that he had a strange experience behind his cabin there in May, 1997. The Kristoviches are, and have always been, the only permanent family at Tombstone Bay—a picturesque spot on the side of the steep-sloped fjord. The forest there is dense and impenetrable, and bears and goats are their only neighbors in the rugged mountains on either side of the long inlet. He had been the only person staying there for a while. Late one afternoon, while digging in his garden by his cabin, Mr. Kristovich heard something strange.

"I heard a terrible yell, louder than a person could make, from the forest, and louder than any bear could possibly make. I have trapped and run into brown bears all my life. Not a bear. The slopes around the cabin were logged for 500 yards around and it had to be coming from beyond that. The first yell was about fifteen seconds long and it repeated itself two more times over the next minute. It was like a scream, a yell and a roar all in one. To me it sounded like it was in a lot of pain. I know it was not a bear in pain. Those yells had quite a deep pitch to them as well. In all my years up there I never heard anything even close to that. I took my rifle and went out there and looked around but couldn't see anything, even tracks!" [V97A-map6]

- 1998: Rob Shelton, Ketchikan, reported in an interview in 2000 that he had heard unusually loud screams at night from a secluded spot on Betton Island, just a few miles west of the southwest tip of Revilla' Island.

"It was March and I had gone camping by boat with friends on the east side of Betton Island, sixteen miles east of Ketchikan. There were no other campers there after we had gone to bed. Around midnight, one of the boys and I awoke to a loud scream and stood up to listen. It was coming from the timbered point to the south of us, a good ways off. It was real loud and lasted about ten seconds. It kind of sounded like the cry of a boy in pain but louder.

"Just a few seconds later it came again, shorter this time, about five or six seconds long, but just as loud and clear. We had a big rifle and had nothing to worry about and after a while we went back to sleep. The next morning we went down there to the point. We tried to bushwhack into the real dense brush and timber where we heard the sound coming from but it was just too thick.

"We had heard no rifle shots at any time the day before. I've heard owls, but that was no owl! Also, there was no sign of people having landed or camped anywhere along there recently. It certainly wasn't a bear. In twenty-five years in the bush, I've never heard anything even close to that sound!" [V98A-map4]

• 1998: Ms. S. L. is a successful Ketchikan businesswoman, who is able to take regular getaways on her boat to remote wilderness spots around Revilla Island. She reported in late 1998 that when she was anchored close to shore on the west side of Thorne Arm on Revilla Island in May of that year, she and her companion heard a single, loud scream from the trees, "louder than a bear could possibly make." [V98C-map5] She also stated that while anchored at the float at the head of Rudyerd Bay, Misty Fjords National Monument, AK, in July of that year, they also heard a long series of manlike yells, "but very loud and angry sounding, like whatever it was didn't want us there!" [V98B-map3]

• 1998: Sharon B., a Ketchikan outdoorswoman I have known for some years, reported that she and her boyfriend Ken had returned to their annual deer hunting campsite several miles west of Thorne Bay, Prince of Wales Island, on Oct. 12, 1998. The second night there they were up late, some time past midnight, when they both heard a loud scream from the forest past their picnic table. Sharon listened through the truck window as she watched her boyfriend head back to the truck, from thirty yards away.

"We were alone in the campground except for one oldtimer who was camping out of his truck on the other side of the campground. It was loud and scary. So much so that as Ken was hurrying back, I was starting to reach for the rifle in case I needed to cover him.

"The night before, around midnight, I had a funny feeling that I never get. It was peculiar and made my hair stand up. After wondering awhile, I just hit the sack. Next day, Ken told me that about the time of the scream he thought he heard a strange sound, like a branch being struck against a tree. We had really come to enjoy the peaceful beauty of this little spot the year before and had enjoyed deer hunting out of there. But this really made me sad because it ruined it for us. We won't go back." [V98C-map4]

• 1998: Mr. M. D., a Ketchikan hunter, reported to me in 1999 that early in November of 1998 he and two other men had been twenty miles into the mountains northwest of Ketchikan, Revilla' Island, when they heard something odd.

"We were set for deer hunting out of my truck camper at a clearcut at the end of Brown Mountain Road. This was in a combination of mature hemlock and second-growth forest at 2,000 feet in elevation. At about midnight, we were all awakened by a long roaring scream from the trees a hundred yards away. It was louder than anything I've ever heard. I've worked on jets. This was loud! The boys all jumped up. The animal that made it was as loud as a jet engine close by, not a bear. We all decided that we should move the camper closer to the main road and drove down a couple of miles. We got a deer the next morning down below and didn't hear the sound again." [V98E-map4]

• 1998: Mr. Jake Lauth, a Ketchikan prospector, reported that on November 20 he had been working one of his mineral claims at Sea Level Mine on the east side of Thorne Arm, an inlet on the south side of Revilla' Island, when he had heard some strange noises.

"I was up a road about five miles from my boat anchoring south of Elf Point on the northeast side of Thorne Arm. The claim was a large open area of blasted rock beside which I had just set off an explosion. After the dust had settled and all was quiet I had started in on turning over the samples. Then from a long ways off I heard the sound of something stepping on a log and snapping it. It was something real heavy, so I loosened my sidearm and hunkered out of sight below a length of large tree trunk which had fallen onto the edge of the claim. I was in a rock depression and out of sight to anybody.

"After a while I heard the same sound again only this time much closer. It got quiet again and then, just as I was just getting ready to get back to work again, I heard a real unusual sound, something a bear wouldn't make. It was the sound of two fair-sized rocks being picked up and 'clacked' together. I waited a bit after that; I didn't see a thing although I sure kept an eye out. [V98D-map5]

"What disturbed me was that after my next trip back, I discovered that something had gotten into my demolitions box, but without pry marks or scratches. It is a large chest freezer kept locked, and something had twisted the lid right off its hinges without leaving any marks. I don't keep any food in there that would attract a bear, and nothing was taken."

• 1998: In an interview in 1999, Ms. Ray S., Ketchikan, reported the following second-hand account involving an incident of "log-rapping."

"My friend S. told me she was with her boyfriend late one November night in 1998 near the top of the Brown Mountain Road [north of Ward Lake near Ketchikan on Revilla' Island] to watch for a meteor shower. She said they had parked their truck on a level layby where the road crosses the last clearcut to the right in the valley below. The snow at the time would not let them get beyond that point. They were enjoying watching the meteors in the night sky while laying on the hood of their four by four truck, which was quite warm and comfortable. The night was cold and clear and the visibility was excellent. Everything was quiet.

"Suddenly there was a big crashing in the woods above them and a real loud "boom, boom, boom." She said her girlfriend then told her, 'We sat up. The hair on the back of our necks was rising. The next moment we heard a real loud "thump, thump, thump" like wood beating on wood. We didn't wait to hear anything more. We got in the truck and headed down the mountain.'" [V98F-map5]

• 1999: Hunter Bob E., Ketchikan, reported in 2000 that August 1,

1999, while deer hunting one-quarter mile south of the east end of Gertrude Lake, twelve miles east of Hydaburg, he and companions had stopped to listen to a most unusual call. He was with two other men who had flown into the Forest Service cabin at Josephine Lake, and were hunting the ridge between Gertrude Lake and Green Monster Mountain at the 2,600 foot level when they heard it.

"It was louder and longer than an elk call, about ten seconds long, and had a deeper ring to it. I know there are elk reported on Prince of Wales, but I grew up on elk down south and am very familiar with their calls. It only came once, really long and loud from the forest." [V98G-map4]

• 2000: Dave Timmerman, a Ketchikan man, was on Brown Mountain for another loud vocalization event, in the spring of 2000. He stated to me in June, 2001, that he had driven a ways up the Brown Mountain Road in April to gather cedar bark for family weaving at a viewpoint overlooking Brown Mountain and its steep valley below. From the alpine and forested slopes across the valley came a haunting yell, not wolf nor man, something louder and from further away. "It was too loud deep and long to be either," said Dave, "We all know what wolves sound like here, it was definitely not that. And it wasn't quite human either and too loud as well. There was something else there as well, a tree with a limb all freshly debarked. It was horizontal, but too high off the ground for any bear to reach from the ground, no claw, tooth or beetle marks, no sign of any tools. I thought it had been a bear at first, but whatever did it would have to be stronger than a bear, it was really a fair-sized tree." [V00A-map5]

• 2000: A Ketchikan woman reported to me that in early December, 2000, she heard two loud roaring screams outside her remote trailer late at night, in the Whipple Creek area north of Ketchikan. Each scream lasted about eight seconds and came from the trees twenty yards from the trailer. She added that the first scream had coincided with a power blackout at the trailer and all the interior and yard lights had just shut off. The second call came only about twenty seconds later, but seemed to come from ninety degrees to left of the first call. Where the trees started, this new direction was about fifty yards apart from the direction of the first call and, to her, suggested that the caller was moving quickly. [V00B-map5]

Jimmy McK., a Ketchikan teenager, had coincidentally reported a fifteen-inch five-toed track 100 feet from the same trailer the previous July, along with a number of seven-inch bear tracks. The larger track, "clearly showed a big first toe and a wide heel," he said, "and did not look like the other bear tracks at all."

• 2001: Jack Stockli, an experienced Ketchikan hunter, told me that in June, 2001, he had been camping with a friend on the beach at Exchange Cove, near the northeast tip of Prince of Wales Island when they had experienced an unusual event.

"We had been sleeping in our tent, when we heard a loud cry from nearby. It was the strangest sound, a loud, rapid chattering and chuckling sound, lasting about eight seconds. Then nothing. It wasn't a grouse or any other animal I've ever heard in the bush. It was the strangest sound. In the morning, there were no sharp tracks around the tent, and it left no clue what it was." [V01A-map]

• 2001: Kenny Smith, a Wrangell man, reported some interesting vocalizations he heard while deer hunting with his friends on Zarembo Island in November, 2001. In an interview in 2002, Kenny stated:

"We had been on the north side of Zarembo, about half way between St. John's Bay and Little Bay and we had got a deer early that day by some muskegs up off the main road. I had gone back about three hours later and had hiked up off the main road through the muskegs to a spot about one quarter mile from where we had got a deer earlier. I was just sitting down and had started blowing a deer call, when I got heard a strange sound from a ways off. It was a loud screaming, howling, roaring kind of sound, about four seconds long. Now I have hunted elk around here and northern California and it was definitely not an elk, a bear or a wolf. It made my hair go up!

"I blew the deer call two more times, every two or three minutes. Each time that sound would answer, each time louder, or closer, I couldn't tell which. But after the third time—I can tell you—I decided to pack it in. Meeting up with the guys, we talked it over. We all agreed it couldn't have been a mountain lion, even if there had been one on Zarembo. The sound had a guttural quality to it, and was too loud. I don't know what it was. The locals have stories about *kushtakaas*, of course, and there are some guys who have stories about seeing things from around here, but I only know that sound was unlike anything I have ever heard before." [V01C-map2]

Kenny admitted that apparently word of his hearing the strange call had gotten around and he had evidently become the target of some good-natured ribbing about it, but was confident about what it wasn't, i.e., not an elk, or any of the other animals occasionally heard in southeast Alaska. When he heard it answering his deer call, Ken added that the thought of an owl had briefly crossed his mind, but ruled owls out because of the call's loudness, length and quality. It may be noted that the mating wail of a cougar would not likely be heard as a response to a deer call. Again it might be noted that the scream was heard three times, a frequency of vocalization not uncommonly attributed to sasquatches.

To ascribe all of these vocalizations and such to sasquatches may at first seem highly speculative. But to suddenly invent other new mammals or birds to account for them is equally tenuous. Where, then, does that leave us? When we compare the types of vocalizations heard with reports of alleged sasquatch vocalizations from elsewhere, it is noteworthy that all these different sounds are already previously documented by other authors as possible sasquatch vocalizations. Of course, they may all belong to sasquatches as easily as to any other "unknown" animal. Until more documented data and recordings are forthcoming, it seems a workable hypothesis to place such calls within the possible spectrum of vocalizations for sasquatches.

The volume of screams attributed to sasquatches have been likened to the sound of a jet engine or louder than an elephant trumpeting. These would both be more than 120 decibels. What else could generate the sounds people are reporting? Bears in pain and cougars are the alternative candidates for loud screaming, but not calls sustained for ten seconds. On the islands in Alaska, cougars can almost always be excluded for zoogeographic reasons, while a bear screaming in pain is not usually heard without a rifle shot or a porcupine handy to implant a few quills. Bears may scream when initially wounded or trapped, but subsequently moan, cry or are silent. In addition to the rarity of their screaming, bears do not usually generate the decibels of the sort generally purported here to be sasquatch vocalizations, nor do their screams carry the distance or sustain a steady pitch for the duration often heard in many of the reports. Whooping and humanlike yells of great volume would also be especially difficult to ascribe to any other regular southeast Alaskan mammal or bird.

Occasionally owls will call, whistle or hoot in a way that might be mistaken for a sasquatch, but their volume is not significant compared with reported screams. Alaskan bird calls are completely documented in a two-CD set entitled *Bird Songs of Alaska*, compiled by Leonard J. Peyton and available through the Cornell Laboratory of Ornithology.[5]

Humanlike yells of short duration attributed to sasquatches are also reported but often with mention made that the volume or carrying power was very great. Cries akin to those of a woman or baby crying attributed to sasquatches in Alaska and B.C. are noted less frequently, as is chattering or chuckling. Bears are able to make a sound much like a child sobbing and may at times have been mistaken for a sasquatch. Ptarmigan (found at higher elevations) and porcupines (mainland only) are both capable of making chuckling sounds but not with any great volume. Some traditional Native narratives ascribe the ability of sasquatches to mimic a variety of calls made by jays and ravens. This is certainly possible, but at present I do not know of anyone who has witnessed the animal making such calls. This is also the case with rapping, tree striking or chest beating allegedly attributed to sasquatches, although a large primate such as the sasquatch might easily be expected to indulge in this behavior within the cover of forest, especially if it were threatened by man's presence. Rapping or wood-chopping sounds in the absence of such human activity is not only noted in many Native narratives along the West Coast, but it has been documented in close association with a number of sightings from California to Alaska. It seems likely that this is a locator behavior or possibly a territorial or threat display.

Outside Alaska, one report has produced a tape recording that has been subjected to scientific analysis. In 1977, Alan Berry, a Sacramento journalist, made available to sound engineers a remarkable tape recording that he stated he made on the night of October

21, 1972, at a bigfoot research field camp at 8,200 feet elevation in the High Sierras of northern California. With researcher Warren Johnson, Berry had been able to remain at a small campsite where repeated vocal and track evidence of sasquatches had been witnessed over several seasons.

In the spring of 1977, an electrical engineer, R. L. Kirlin, and a University of Wyoming graduate student, Lasse Hertel, initiated detailed research on the tapes using state of the art signal processing. Presented as a paper in 1978 at a symposium held at the University of British Columbia's Museum of Anthropology, and subsequently published in Marjorie Halpin and Michael Ame's book *Manlike Monsters on Trial—Early Records and Modern Evidence*, the engineers delivered their summary as follows.

"Having analyzed a tape recording of purported Bigfoot speech using accepted techniques of signal processing, the authors conclude that the means and the ranges of the recorded pitch and estimated vocal tract length of the speakers indicate that the sounds were made by a creature with 'vocal features corresponding to a larger physical size than man.' They also conclude that the tape shows none of the expected signs of being prerecorded or rerecorded at altered speed and hence diminish the probability of a hoax."[6]

In his book *Bigfoot*, author and researcher Alan Berry[7] describes the High Sierra events in detail. The recordings were lengthy and astounding, containing grunts and chattering in strings of vowels and consonants, which I believe are highly suggestive of simple speech. The acoustic analysis indicates a biological origin and is publicly available.[8]

Kirlin and Hertel also noted, "In addition, there is a wide range of vocalization, much of which shows a human-like level of articulation. There are also considerable lengths of what might be termed moans, whines, growls, grunts, and even some whistles, which no primates other than man are known to produce. The phrase might be written, 'Gob-uh-gob-uh-gob, ugh, muy tail...' The results indicate more than one speaker, one or more of which is of larger physical size than an average human adult male.... The average vocal tract length [of the largest creature] was found to be 20.2 cm. This is significantly longer than for a normal human male. Extrapolation of average estimators, using human proportions, gives height estimates of between 6'4" and 8'2". Also, the sound /g/ in 'gob' suggests a humanlike vocal tract [two vocal cavities]."[9] Other professionals have listened to the tapes and have expressed their opinions, which have essentially been qualitative."[10]

If we consider the possibility that apart from frequently reported apelike screams, hoots, grunts, roars and growls, there exist, as reported in Alaska, B.C. and elsewhere, the less often heard whistling and chattering (with elements of what linguists might call phonemes—"our" consonants, diphthongs and vowels) then we are left basically with an ape that has a much greater potential for communication within its species, if not simply between mother and offspring.

This does not necessarily mean that sasquatch mothers would be able to vocally identify the hundreds of botanical species within their varied ecosystems, although that would certainly be useful. But if they are intelligent enough to recognize the differences between edible and poisonous species, as apparently are other apes or bears for that matter, then there is just the possibility that a sasquatch might have simple vocalizations for such or at least for simple constructs as "good" or "bad." And while perhaps sasquatches do not have a species-wide "language," they may perhaps at least use a system of "naming" within a mother-offspring family unit. Of course, this is purely conjectural. Perhaps there is no meaningful significance to these reports of rich chattering and the sounds are simply expressive of various emotional states.

Whatever the significance of speechlike chattering to sasquatches, they have certainly seemed to perplex Albert Ostman, Alan Berry, Warren Thompson, Bruce Johnstone Sr. and others who have reported them. It is not the humanlike mixture of the sounds, sometimes reported as a kind of "speech" or "language" that is so puzzling. It is rather the way in which these reports suggest a "dialogue" between the creatures themselves and the people who call back and forth with it.

Vocalizations leave a lot more questions than they provide answers. If most reported human-sasquatch interactions involve people wandering into a territory defended by a sasquatch, with all the threat displays, yelling and screaming likely to occur, then such investigators are just as unlikely to be privy to any of quieter, possibly more meaningful sasquatch dialogue as occasionally reported. It is interesting to note however, that in all the above such reports of back and forth communication, the people involved were reportedly not wandering around at the time, but had already been settled down peacefully in some sort of camp. This would seem to be a favorable way to initiate a more peaceful interaction with such primates, having given any hypothetical sasquatches involved more opportunity to first observe the nature of humans as peaceful and nonthreatening.

Perhaps this is why in the majority of vocalization reports there is no sighting, in fact, often only the sensation that one is being watched. The human reaction to this rises on a scale to the point of wishing strongly to leave the area, often with all of the associated sympathetic nervous system reactions, such as piloerection (hair standing on end), perspiration, etc. The autonomic nervous system is capable of doing this at the brainstem level, without one really even knowing what one may have seen, smelled or heard at a subliminal level to produce this protective avoidance behavior. It has been reported not only by hikers and campers, but as well by veteran bear hunters, who seem unable to explain why they, well familiar with the odor of bears, should also have such a reaction.

Popularly referred to by sasquatch field researchers as "the pheromone theory," this interesting fear response or sensation of "being watched" is something I have numerous reports of from Alaska. Briefly stated, it is the human body's unconscious response to a stimulus, in this case an odor, resulting in anxiety, hair erection and a general "flight or fright" reaction, before the brain can correctly identify the cause. The number of scent molecules is strong enough to trigger an increase, through the lower brain, in adrenalin, but not enough to register on the higher centers in the brain. The fact that it may possibly be just as easily elicited by the subliminal presence of bears, however, causes me to withhold such reports when the fear response is the only thing noted. It does strike one as odd that many bear hunters spend a lifetime hunting without ever experiencing this unusual reaction. It may be that there are far more near encounters with sasquatches narrowly averted by this type of reflex reaction. We really have no way of knowing. Mutual avoidance would certainly have survival value to the two species—one the possessor of great strength, the other the possessor of great intellect, greater numbers and sharp projectiles. Perhaps it is just as well that the individuals reporting such puzzling sensations have not questioned the wherefore and the why of their "rising hackles," relying instead on the possibly ancient wisdom of their autonomic nervous systems.

Noisemaking, such as tree striking (often called "rapping") or log striking with a peculiar wood-on-wood quality, has been noted by many researchers and authors. Its purpose is not clear, but based on similar behavior in chimpanzees, it may be a locator behavior, related to finding another sasquatch, defining territorial boundaries or perhaps simply a part of a repertoire of a threat display. Rock clacking is also noted infrequently and may have a similar function.

Whatever the vocalization or noisemaking attributed to sasquatches, they are not usually seen making the sound. This would argue that they simply don't want to be seen or heard, at least by non-sasquatches. In the dense forests of the West Coast and elsewhere, these sounds may be as close as any of us might ever get to a sasquatch—but, as reported by those hearing them, they're not likely to forget.

1. Bindernagel, John, 1998, *North America's Great Ape: the Sasquatch*, Beachcomber Books, Courtenay, B.C. pp. 191–194.

2. Green, 1978, *Sasquatch: The Apes Among us,* Hancock House Publishers, Surrey, B.C. pp. 386–395.

3. Green, ibid., p. 389.

4. Jackson, W. H., 1974, *Handlogger,* Alaska Northwest Publishing Company, Anchorage, pp. 156–158.

5. Peyton, Leonard J., 1999, *Bird Songs of Alaska,* Library of Natural Sounds, Cornell Laboratory of Ornithology, Ithaca, New York.

6. Kirlin, R. L., and L. Hertel, 1980, "Estimates of Pitch and Vocal Tract Length from Recorded Vocalizations of Purported Bigfoot," in *Manlike Monsters on Trial,* eds. Halpin, Marjorie M., and Ames, Michael, University of British Columbia Press, Vancouver, Canada, p. 274.

7. Berry, Alan and Barbara-Ann Slate, 1976, *Bigfoot*, Bantam Books, NY.

8. Berry, Alan, 1998, *The Bigfoot Recordings,* Sierra Sounds (CD and cassette), P.O.B. 2248, Mariposa, California.

9. Berry, 1976, pp. 287–288.

10. Berry, 1976, p. 274.

13 Strange Encounters in Southeast Alaska

Flying Rocks and Shaking Cars

Sasquatches seem to be regularly reported as responsible for throwing rocks, branches and logs, as well as smashing trees. These alleged behaviors are frequently mentioned in other literature, occasionally in association with a reported sasquatch sighting or unusual screams. The threat displays of chimpanzees and gorillas both include breaking and throwing branches. It is not too difficult to envision this behavior as a part of a medium aggression display by sasquatches. Bindernagel[1] puts forward an analysis of various levels of aggression in sasquatch behavior, implying a rising scale of belligerence related to an apparent ordinate increase in the anxiety of the individual ape. In the case of reported sasquatch behavior, he notes the following sequence of threat displays: benign stone throwing (mild deterrence?) followed by aimed throwing of large stones or chunks of wood (intimidation?). As well as displaying openly aggressive or defensive behavior while brandishing sticks or branches. Some reports of wilderness "vandalism" describe feats requiring prodigious strength. It may be that these "deterrent" types of behavior are simply the work of a disgruntled sasquatch. The following is a brief chronological listing of purported sasquatch behavior toward man and wildlife, gathered from southeast Alaska.

• Mid-1940s: A seventy-two-year-old Ketchikan woman told me in 2000 that around the time she was six, she remembers accompanying her family up a deer trail on Prince of Wales Island. She was not certain of the exact location, but thought it might have been on the island's south tip, west of Point Chacon. Her family would regularly go there for gathering berries and seafood.

"On this particular summer day, we had been all together up a deer trail beside these thick bushes, when all of a sudden we saw this four- or five-foot log, about ten inches in diameter, come sailing into our midst and bounce to a stop among us on the open trail. It had been thrown in an arc with some force, but the strange thing was that it had been thrown without a sound. If a bear had smacked it you would have heard something. And we were the only people up there. I remember asking my parents about it on a number of occasions for years after, but all they would say was, 'No, that wasn't a bear.' But when I would ask them what it was, if it wasn't a bear they would just look at me and say, 'Well, there are some things in this world which are just not well understood,' that's all." [B40A-map3]

It is not clear who or what threw the log, but by the mention of silent throwing and the absence of other humans in the area, there are not really many candidates besides a hypothetical sasquatch. Although logs and rocks are the most usual projectiles alleged to have been thrown at people in the bush, they are not the only objects reported, as noted by some Alaskans.

• Late 1940s: Longtime "southeast" resident, Mrs. J. M. of Ketchikan, confided in an interview in 2000 that her son-in-law and his brother had been trapping out of a boat in the vicinity of Thomas Bay, on the mainland north of Petersburg, sometime in the mid- to late 1940s, when they had an unsettling encounter with something strange.

"They had a cabin cruiser and had anchored up just yards offshore in Thomas Bay to escape a snowstorm. There were whiteout conditions and they were about twenty-five yards or so from shore in a sheltered spot. It was getting quite dark out and the men had gone to bed. Suddenly huge snowballs began pelting their boat. As the one brother stood outside looking at the shoreline he yelled out to shore, 'Heyyy, cut it out!' It was then that he noticed with some alarm that the snowballs hitting the boat were at least eight or nine inches in diameter. He later showed me with his hands just the size the snowballs had been. Right then, he knew it could not be the kind of snowball a human might make, and they got right out of there.

They kept quiet about the incident, as they thought no one would believe them and did not want the ridicule." [B40B-map2]

Although only one person I know of outside Mrs. M's family has apparently heard the men's fabulous story, giant snowballs are about the last thing one might include in any fabrication if believability was the goal. Yet, astonishingly, there is a recent report of a giant snowball being thrown at a couple parked in a remote snowy spot at the end of the road, Harriet Hunt Lake, Revilla Island, in early winter, 1997. The couple did not see anyone between them and the nearest trees, about ninety feet away. As in the previous account, the people involved reported leaving the area immediately, as the snowball was "about ten inches in diameter, and came out of the trees ninety feet away in a low arc." [B79A-map5] On investigating the case in 2001, I noted the distance to the nearest trees was indeed ninety feet. The trees, however, were slightly downhill from the parking spot and whatever may have launched such a snowball from the concealment of the nearest trees would certainly have required as much strength as an Olympic hammer thrower.

- 1950s: A Miss M., now living in Ketchikan but then residing in Klawock, Prince of Wales Island, and others reported damage to the metal roof of a Klawock house in conjunction with house shaking, screams and sightings of hairy hominids on the south edge of town right beside Klawock Lake. As mentioned in chapter 5, according to several people interviewed, bear traps set overnight for the creature(s) were found in the morning sprung with sticks. [S51A-map5]

- Late 1940s or early 1950s: An old report through Mr. A. K. and his cousin, the late W. K., both of Ketchikan, came to my attention twice in 1996. A. K. told of "a caterpillar tractor tipped over on the job near Hollis, Prince of Wales Island, two nights in a row. My cousin said it took the logging crew, who certainly must have had more productive and rewarding things to do, the better part of each day to right the tractor." There is one other unconfirmed report, in the same area and at about that time, of a large caterpillar tractor found with both tracks broken off. [B40C-map4]

- 1966: A Ketchikan woman known to me personally, "Doris Yeoman" (not her real name), recounted a frightening episode that occurred to her and five friends involving something that allegedly lifted their car one night at a recreation spot just north of Ketchikan.

"It was an August evening in 1966 and we were enjoying an evening at Ward Lake. There were six of us, and we were all around the car, a big old Ford Thunderbird, parked in the old parking lot at Ward Lake, close to the water and only about thirty feet from the forest. The car was parked facing the exit to the road out. There were three guys and three of us girls. It was cloudy, but not raining and it was about 11:00 p.m., dark out. It was real quiet; there was no wind at all. We were the only ones there, just hanging out around by the beach. But one of the girls heard some noise in the bush and we all scrambled back in the car. The guy who was driving started the engine, but before he could turn on the lights or put it in gear something picked up the whole rear end of the car off the ground. One of my girlfriends screamed and the driver gunned the engine. But the wheels were off the ground, and we had to just sit there, wheels spinning in the air. By the angle of the car I would say we were tipped forward at least thirty degrees. The guys were yelling at each other to go, and the girls were both screaming. I just put my hands over my ears to block it out.

"Then, whatever it was that was holding us up just started shaking the car from side to side, not letting it down at all. It just picked up the back of the car and shook it like a toy. The car had its wheels in gear and spinning the whole time. It was shaking like that for over a minute, maybe a minute and a half. And then whatever it was just dropped us. The wheels spun on the gravel and we were out of there. Later on, one of the girls said she had seen something large and brown standing in the bush. We didn't go back there for a long time, and now all but one of the others have moved away." [B66A-map5]

Bindernagel's research documents eight other reports of sasquatches shaking vehicles, including trucks and travel trailers.[2] This report from Ward Lake was one of six reported sightings within one mile of each other, with an additional two other sightings, one report of extremely loud screaming, and two nest reports within a mile and a half north along Connell Creek and the Connell Lake Road. There are an additional two curious reports, within weeks of each other in late September and early October, 1999, of the remains of an eight- or nine-inch foot with black hair, "not a bear's," examined beside Connell Creek, less than a mile from the Ward Lake

Figure 36: Author's impression of sasquatch reported shaking five people in a car at Ward Lake, Revilla' Island, AK, in 1974 as reported by witness L. W.

parking lot. These reports extend from 1960 to 1999, with half of the reports describing a brown creature and the other half merely a "dark-colored creature." It seems that the combination of reports from this area of old-growth, salmon spawning streams and mountainous forest may in fact provide an ideal habitat for one or more sasquatches.

• 1974: In a most amazing follow-up report to the last, at exactly the same spot north of Ketchikan, but separated by eight years, Ms. L. W. of Ketchikan had the following report to put forward.

"In the first week of June, 1974, I went with four classmates, all friends of mine, to Ward Lake to just take it easy and enjoy being out of school. Dan and Webb were driving a gold-colored four-door Dodge Dart, and Deb, Cindy and I had gone with them. We were all chums and classmates. That evening we had just decided to cruise around up to the parking lot at Ward Lake. It was about 10:30 p.m., and we had parked on the gravel turnaround a few feet from the trees and lake. We had the car off; it was dark out, nice and quiet and were just talking. All three of us girls were in the back seat. All of a sudden we felt the back of the car start to bounce up and down.

"'It's a bear! It's a bear!' the boys were calling out. One girl screamed and the boys took a look in the rearview mirror.

"'No!' one of the boys called out, 'It's a *kushtakaa*!'

"The bouncing was rhythmic, the rear end of the car going up and down just like something was lifting it up and down every few seconds. The engine was off and the back end of the car was going up and down a foot or more. This went on for maybe two minutes, but it seemed like a lot longer. After it stopped we just stayed in their car for a long time before we got out to take a look around the car. Dan saw some dark hairs on the back of the car and just brushed them off onto the ground, saying, 'Aww, it was just a bear.' But the way the car had been bouncing felt to me like on the upswing, it was being really lifted up each time, not like what you would expect a bear to do." [B74A-map5]

One explanation of the two preceding reports is that the shaking of the two vehicles may have been play behavior of some sort by a male, perhaps an adolescent, sasquatch. Alternatively it may have been gentle territorial behavior by a male or even just curiosity-related behavior. The 1966 report of the vehicle being held well off the ground with the rear wheels spinning clearly does not describe bear behavior, while in the 1974 report, one of the boys is alleged to have stated that when he had looked in the rearview, it was not a bear that he saw, but a "*kushtakaa*."

• 1984: In a previously mentioned report, a Ketchikan woman, stated that while traveling from the Hollis Ferry terminal to Thorne Bay, Prince of Wales Island at night, driving thirty-five miles per hour up a hill, she and witnesses in the car watched a seven-foot-tall creature with long, dark hair keep pace close beside the car for about ten seconds. [S84A-map4]

• 1988: Mrs. Brenda R., Ketchikan, recounted seeing damage to a neighbor's car, summertime, around 1988, as noted in the sightings chapter. The neighboring family was just returning to Ketchikan from camping at Ward Lake, Revilla, Island, and they were discussing a sighting of a large creature which walked on two legs

Figure 37: Illustration depicting encounter south of Hydaburg, Prince of Wales Island, AK, in 1987, where two hunters reported being assailed with large rocks as they were moving their deer out of the forest to their boat. Sketch by author. [B87A-map4]

across the parking lot and past their car, scratching the side of the vehicle in doing so. The witness observed four or five long, parallel scratches down the passenger side of the vehicle. [B87B-map5]

• 1988: Mr. V. S., a Klawock commercial fisherman, reported to Al Jackson and others in 1996 that while anchored on his forty-foot commercial boat in Affleck Channel north of Prince of Wales Island one night in summer, 1988. "I and my crewman, W. D., both felt the boat suddenly list. As we were getting up from the galley, we heard the sounds of heavy footsteps running quickly forward on the other side of the pilothouse. As we ran outside, around the back of the pilothouse, to shine our flashlights forward on whatever, we felt the

Figure 38: Photograph of dynamite storage freezer with lid, lock and hinges torn off, at Sea Level Mine, northwest Thorne Arm, Revilla' Island, AK. This account was reported in 1998 by prospector Jake Lauth.   Photo: Jake Lauth

boat rock and heard a loud splash in the water off the bow. We could see no sign of anything around the boat." [B88A-map2]

In another late eighties report to Al Jackson, a surprised Klawock man stated that in the fall, around 1986 or 1987, he and two other man had been deer hunting north of Klawock and had got a deer on shore near their boat late in the day. It was getting dark and it had taken them a while to get the deer down to the water. "We had just got to within sight of the boat and were finishing cleaning it out, when all of a sudden from out of the trees these big rocks the size of basketballs started flying at us! We didn't hang around to see who was doing the throwing. We just got the deer up off the ground and ran with it to the boat!" [B87A-map4]

Some confirmation of this recently came from a Hydaburg man, who in 2001, told me an identical story, but added: "My father had been just off the beach the whole time, running the boat, heard the rocks and saw the men come running. 'It was pretty amazing,' he said."

- 1991: There exists a second-hand report through Mr. Larry Thomas, Ketchikan, of a security guard, A. R., who stated to Mr. Thomas in 1992 that in the previous year, at the place of Thomas' 1992 sighting near Ward Lake, he had been knocked unconscious by a rock thrown through his vehicle window from out of the dense roadside forest. A. R. told Larry that he "did not think it was vandals." He also noted his dog behaving very strangely, which was curious to him, "because the dog was used to bears." [B91A-map5]

- 1996: As previously noted, Mr. L. D. reported that he and his uncle, after hearing screams while deer hunting near Bostwick Inlet, south Gravina Island, found some of their camp supplies rearranged. [V96A-map5]

- 1998: Jake Lauth of Ketchikan, a part-time prospector and miner, reported to me in 1998 that besides damage to his tent camp on the east side of Thorne Arm, Revilla Island, and unusual "rock-cracking" sounds the season before, his outdoor full-size chest freezer (used as a dynamite storage chest) "had been ripped opened the lid twisted badly, being torn off at the hinges, with no claw marks or scratches of any sort on the enamel." Jake indicated "it was not a bear" and that "no tools had been used." There had been "no food in the freezer to attract a bear," and he did not suspect vandalism as no dynamite had been taken. [B98A-map5]

The following throwing behavior, which may correspond to the hypothesis of mild deterrence, was also noted in the account of a Tsimshean hunter who in a report from Duke Island given in 1992, recounted in the chapter on Native narratives, experienced pine cones thrown at his back as he was sitting at his campfire. An older Tlingit tale from north of Sitka, recorded in 1909 by Swanton, gives a similar account of a *kushtakaa* in a tree doing much the same thing to hunters on the ground below. In Swanton's tale, however, rocks were initially thrown and both the "sasquatch" and the hunters downplayed the interaction by substituting pine cones. This last folk tale also has one or two additional motifs that should probably be interpreted as recurrent stylization in Tlingit folklore, such as the notion that breathing the creature's ashes would cause convulsions. This folk tale should be considered only as a possible allusion to documented sasquatch behavior.

A British Columbia university archaeologist similarly reported to me in 1979 that while on an early spring camping trip to Cultus Lake, B.C., around the early seventies, he and a lady friend had spent an uncomfortable night in a tent there experiencing much the same thing. He told me that "all night long, our tent, which was in a secluded, hopefully romantic spot just outside the perimeter of the main campsite, well deserted and not under the lee of any cliff face, had been bombarded with a continuous shower of small stones which were being lobbed in arcs onto our tent. There was no laughter or any sign of a prank, and the very length of time it went on suggested to me that we were not the playful target of any human being. It really ruined our reason for being there."

It is possible that the throwing of small stones or pine cones by sasquatches is, as put forward by Bindernagel, a form of mild deterrence. However, it may also simply be a form of play. In no case of throwing that I have heard of were vocalizations, angry or otherwise, ever reported by the witnesses.

There is another Ketchikan report involving pine cones, secondhand from several people that I have not been able to fully confirm. Bindernagel lists at least one report in British Columbia of pine cones allegedly thrown by a sasquatch, and this has been mentioned as well in Tlingit folklore.

Dave and Alberta M. of Ketchikan related to me in 2001 what appeared to be a case of a sasquatch "caught in the act." The couple really enjoy the scenery and tranquility that the forest roads of Revilla' Island offer, enjoy seeing wildlife, and don't let stories of *kushtakaas* stop them from getting out. On this occasion, however, something was to spoil their recreation. They related that they were parked in their car near Ward Lake the night of December 23, 2000, when they noticed something "trying to sneak up" on them.

"My wife and I were parked just off Revilla Road, near the junction of with the paved Ward Lake Road, just enjoying the music from the car radio. It was late, real dark, and we had the car running with the headlights shining on the snow in an open area beside the highway. The light was bright off the snow in a big arc to the left and right, and the snow reflected well enough that you could see if anyone or anything were approaching on either side. We had the windows down and it was real quiet out, no other cars or noises around at all. Just then, I noticed a real rank odor, kind of like the smell of wet dog hair, and I looked around. Off to the left, slightly behind the car and to the side, about sixty yards away, I noticed movement, and could make out a large black form, crawling just like a man on his belly. When I looked at it, it would stop. And when I would look back after a few seconds, it would be crawling, and then stop again while I stared at it.

"It wasn't a bear, and I couldn't think what a person would be doing out there crawling belly down in the snow. I told Alberta to roll up the window, I didn't want whatever it was to come reaching up and grab her through the window. In the light reflecting off the snow, I could see what looked like a faint reddish-whitish reflection from its eyes, and the face appeared almost human. I put the car in gear and pulled around away from it and headed back down the highway, but the lights didn't swing in its direction again as we got going. It seemed big, about seven feet if it were standing up, I would guess. I've heard all the stories about *kushtakaas* and that sort of thing, but I've never heard of them sneaking up on a car. But that's what it seemed to us to be doing." [B00C-map5]

It will never be clear what might have happened had the couple not decided to leave. Also, it is remarkable that at the Ward Lake parking lot, less than two minutes drive east, there have been two Alaska reports of occupied cars being picked up by the rear bumper at night and shaken.

In May, 2001, Charles Ryan of Ketchikan gave me the names of four women who were allegedly subjected to the same sort of nocturnal antics while parked recently at Point Davison, the southernmost point on the Annette Island Road system, twenty air-miles south of Ketchikan on a completely different island—one requiring a half-hour boat ride to access. Over the telephone, their subsequent Annette Island report sounded much like the other two on Revilla'. While it is easy to exclude bears and teenaged boys from such herculean feats, one is left wondering why a sasquatch would engage in this sort of repeated behavior. Curiosity, play or a combination seem to be possible explanations, especially if adolescent male sasquatches could be anything like their smaller human counterparts. As well, in all four cases involving parked vehicles, there were human females present in the vehicles and music was playing. In any event, it is also worth noting that no one was reported hurt and, unlike the behavior of adolescent human males, the perpetrators were apparently shy or, at least, nonvocal.

All this points out that with regard to the occasionally proposed "dangerous" or "possibly violent" nature of the sasquatch, the species seems on the whole remarkably shy; if not molested or intruded upon too much it suggests a peaceful disposition. It may be that those who say they are worried about what might happen to them in the bush if they ran into a sasquatch would be better off worrying about that other bipedal primate, *Homo sapien*—not necessarily so retiring, much better armed and more unpredictable.

Missing Deer and Other Stories

Stories of sasquatches have so often mentioned the theft of deer carcasses that, if there is anything to them at all, it could only imply a fondness for venison. In 1948, one such sighting was alleged to have involved a deer being pursued by a sasquatch. As previously documented, Al Jackson has such a story.

"Several years ago, just before he died, my cousin W. K. told me of an experience he had as a teenager. My cousin told me, 'We were all playing on North Tongass Road near Cranberry Road, north of Ketchikan, in the summer of 1948, five of us, when a deer jumped out of the bush on a bluff above the road and ran across the highway like something was after it. It was followed just seconds later by a big black, hairy, ten-foot creature, like a gorilla on two legs, which hesitated for a moment to look at us. It was less than forty feet from the nearest boy, and then continued running across the road into the trees on the ocean side of the road after the deer. My uncle told us not to tell anyone because, he said, "someone might want to shoot it."' [S48 -map5]

• 1989: In an interview in 1998, Ketchikan hunter Steve B. recounted theft of game under unusual circumstances.

"I was hunting on the east side of Carroll Inlet, Revilla' Island, about one and a half miles north of Gem Cove on August 3rd, 1989.

Figure 39: Author's impression of sasquatch chasing deer across road north of Ketchikan, AK, 1948, as alleged by two witnesses and reported by Al Jackson.

Eric K. was with me. We got there by boat and had seen no other boats or signs of hunters—not even a shot. We had come up over a knoll in the muskegs there and I had shot a buck in the chest at 200 yards with a .300-caliber round. He went down hard but we had several little knolls to go up and down over to get to the body. When we got there, just three or so minutes later, there was more than enough blood to have finished him off, but the body was just gone. There were no places to see tracks and there was no blood trail. A bullet like that takes care of a deer almost no matter where it hits, and we saw this animal go down for good. Something took that deer in the few minutes we were out of sight. Sometimes you miss a shot or hit it badly you know, but this deer was hit well, went down well and there was nothing there. It was just like something walked out of the trees back there and made off with it. We looked for hours all round but not a trace. We were both real puzzled. By the lack of signs, I didn't think it could have been a bear that carried the deer off."

The muskeg area north of Gem Cove was coincidentally the location of three humanlike sixteen-inch tracks reported by Ketchikan hunter A. S. in 1979. [B89A-map5][T79B-map5]

- 1990: Sometimes the bodies of deer have been taken with clues as to the identity of the perpetrator. Hunter Robert Durgan reported that in the fall of 1990, or thereabouts, he, his father and uncle were hunting behind Kasaan, east coastal Prince of Wales Island.

  "We had left the carcass of a freshly killed deer a very short distance from the creek where we were washing up. While there, something extremely quiet stole the carcass within the span of a few minutes. The distance from the creek to the carcass was less than fifty yards. We are all familiar with bears and in my opinion, it was not a bear, which would have had to drag the carcass over noisy ground." The hunter stated he and others "had repeatedly seen sixteen-inch tracks in the area." He stated theft of deer carcasses had happened in that area above Kasaan more than once. [B90A-map4]

- 1995: Ketchikan hunter Victor Mulder told me of some strange signs that he had seen one summer not long ago on the east side of Prince of Wales Island. In an interview in 2000, he said the following.

  "I was out with Dan and Jeremy Shelton of California in mid-September, about 1995, up north of Thorne Bay on Prince of Wales Island. It was into the deer season and we were truck camping, hunting up an old logging road, I believe it was near Red Sand Beach. You can get there north of Thorne Bay. We had gone a mile or so up to the end of this road through second-growth and, at the end, had come to a place where the bush opened up. There were alders growing around. It was real quiet, not a sound like you usually hear. In an open area off the end of the road were these four piles of complete deer bones, each skeleton set well apart. Each pile of bones was within a four-foot circle, nothing carried off. All the bones were there. Three of the deer were small 'spikes,' not old at all. They were all surrounded by circles of deer hair just as if they had been plucked. There was not a mark on the bones of any knife or saw, and no signs where a bear or wolf had chewed on any of them. What was really amazing was that at all the four carcasses, every single long bone—you know, the femurs, humerus and all the long bones—had been cracked in half and the marrow had been sucked out! Bears and wolves will spread the bones all over. These carcasses were all in their own little circles, with the hair looking just like it had been plucked out in a larger circle around each one. There was no sign of a fire anywhere; no sign of any human having been camped there that season. No charred bones either. What really got me was the last carcass we looked at. It didn't look more than three to four weeks fresh. I'll say it was enough to give a guy the creeps." [B95A-map4]

This is an unusual report in that, regardless of who or what killed the animals, the bears and wolves should have dispersed the remains or at least made off with the marrow bones, unless of course something else had stayed nearby and taken possession of the kills. The tidiness doesn't seem to fit the behavior of bears or wolves at all. I am aware of humans eating marrow, but the bones in question apparently showed no signs of being chopped, scraped or even boiled. They were snapped. A sasquatch certainly would be able to break open the long bones without leaving tooth marks and wouldn't have to bother with carting them off to the pot either. Although there is no definitive proof, it all suggests that regardless of what caused the animals' deaths, a sasquatch may have had a hand in cleaning up the remains. Of course there are reports of sasquatches chasing deer, deer being carried off with sasquatch tracks in the area to indict them and rocks being thrown at hunters carrying a deer, but unless caught red-handed, it is difficult to prove a sasquatch was responsible.

Elsewhere, sasquatches have been reported seen washing roots (B.C.), stacking them beside a stream (Oregon) and piling up mollusks (B.C. and Alaska). In Alaskan reports where the sasquatch itself was not witnessed, they have been alleged to be responsible for stacks of empty abalone shells and large, neat piles of ripe, wild crabapples. According to retired Oregon Fish and Wildlife officer Jim Hewkin,[3] sasquatches are also attributed (in Washington and Oregon) with hanging deer kills in trees. On this basis it would appear that one sasquatch sign might be the leaving of piles or stacks of discarded food remains. It is speculated this behavior is probably simply a function of convenience after having gathered enough calories to make squatting down for a meal worthwhile, rather than any sort of wilderness "table manners."

By some accounts, sasquatches are capable of phenomenal feats of strength. One such mysterious report was provided by Mrs. Susan H., Ketchikan, who gave a puzzling account of something that, overnight, moved a heavy furnace in the forest edge just behind their home ten miles south of Ketchikan:

"A couple of months ago, sometime in February, 2000, my husband had taken a broken furnace out behind the house with heavy equipment and placed it on a ten-foot flat, level spot thirty feet above the house on the slope going up the mountain. It must have weighed close to 600 pounds. To pick it up you would need at least four guys. It was set back about five feet from the edge of the flat spot, so that if in being moved it fell toward the edge, it would not roll over the edge. The day after he had moved it, we were both surprised to find it was down at the bottom of the slope. There were no marks to indicate it had been rolled over or pushed. As we see it, something must just have lifted it and thrown it over. There are no kids around for vandalism, and we are a long ways out of town." [B00A-map5]

This is a rather nebulous report, and of course there is nothing to disprove human vandalism, but the owners do not think that was a factor. In fact they reported nothing to the police. No beer bottles or any such evidence were found and the house is set along a very sparsely populated strip of highway, in dense forest and up a steep, winding, 150-foot driveway. The house is not visible from the highway. No tracks were found, since it had rained the night before and the rocky ground on the flat spot in the slope was not conducive to imprinting. As we viewed the old furnace and level spot behind the house in April of 2000, we were impressed with the difficulty any person or bear would have in even trying to tip it over. There were also numerous places in the thick hemlock forest above the house where a creature could have sat or stood unobserved while watching Mr. H. working with his heavy equipment to lift the old furnace while landscaping. While there is no clear proof that a sasquatch was responsible, there have been three sasquatch sightings reported since within a mile of Herring Cove, just a few minutes' walk north of the furnace report, all in June or July, 2001.

These reports are all varied, but are consistent with reports of occasionally curious, fearless or even relatively aggressive sasquatch behavior as documented by Green[4] and Bindernagel.[5] It seems that sasquatches do not have any particular fear of man or his artifacts, rather just a general avoidance unless their immediate geographic range is invaded by man, in which case curiosity and, if necessary, threat displays are exhibited. Boats, cars and machinery seem to hold a certain fascination for sasquatches. This would be indicated by the alleged manipulation of the vehicles (four in these

Alaska reports) and perhaps by running alongside a car in the Thorne Bay account given by Mrs. L. A. There are three reports of something happening onboard boats. But in the report from 1979 at Affleck Channel the creature was not seen, only heavy running footsteps were heard on deck, and a loud splash with the rocking of the large boat noted.

There is a fourth very similar boat report I have from a Sitka man, a businessman and commercial fisherman, of a sasquatch witnessed onboard his fishing vessel at night. It was apparently seen picking fish out of a topside hold on the aft deck of the boat, which was not particularly close to land, sometime in the seventies or eighties. The sasquatch was reported to have looked surprised at being "caught in the act" and dived over the side.

Although Green and Bindernagel both document a number of cases where sasquatches have been seen with fish, curiously I do not have more Alaskan reports of this. Perhaps it is because the sasquatch is not slow in disposing of the evidence or perhaps it is because at night people have simply given up being where the fish are, leaving the creeks and rivers to the nocturnal, furred animals who then come out to take their turn. I do not know.

Theft of game attributed to sasquatches is reported a bit more frequently and in the work of other researchers, the theft by a sasquatch is occasionally witnessed. The theft of game is only "theft" if we consider ourselves to be the primary and preeminent inhabitant of an ecosystem. It would appear that from a sasquatch's point of view, this is debatable. "First come, first served," would appear to be the operant principle here. It is also possible that the throwing of rocks, logs or whatever is at hand would not only function as part of a primate threat display, but it might have the intended effect of causing the recipients of the display to leave a snack behind, as is suggested in the 1988 rock-throwing report from Klawock hunters who had to avoid one large thrown rock and run with their deer to their boat.

The careful "hiding" of a black bear carcass in the following Revilla' Island report by Mr. J. B. would appear to match the "theft" of deer carcasses and suggests, if it was in fact a sasquatch responsible, that a certain amount of intelligence went into the caching of the carcass.

- 1998: Mr. J. B., a Ketchikan hunter, reported in 1999 that he and his hunting partner, Mark, had an unusual experience on Revilla Island in the fall of 1998 that raised their hair.

"We had been deer hunting up from a camp between Upper Wolf Lake and Harriet Hunt Lake, north of the Ketchikan area on southeast Revilla Island. It was real late in the day and we were headed back with Mark in the lead and Mark came between a black bear and her cubs. In this situation my partner had only one round left and he had to use it but called out to me to come quick, as the shot had only wounded her in the spine.

"I got one shot in, but the light was bad and she bolted for a small dense thicket of trees, kind of a small island in the muskeg. The cubs were gone; she had lost a lot of blood, but it was getting real dark and we just had lighter deer calibers. Although we felt bad, we knew she would not last the night but would stay in there. We agreed it was probably the wisest thing to just come back real early the next morning. The next morning we were there right before dawn and you could see by the amount of blood on the ground that she was either in there or couldn't be more than a few feet away somewhere.

"Well, she wasn't in the thicket and we couldn't see a body anywhere around on the muskeg although we walked over every inch and there was lot of blood right there. It was as if she had been spirited away.

"After an hour or so of looking, I was just standing there thinking, on a level piece of muskeg when something caught my eye on the ground by my feet. It was showing through a four-inch square opening in the sphagnum and moss. We knelt down and discovered that an area of ground the size of the bear had been neatly covered over piece by piece with patches of moss, laid edge to edge on top of the bear. She was there dead and looked as if she had been placed there in a hollow depression or hole in the muskeg. It looked just as if she had been camouflaged, neatly packed in there, all except one small patch left showing to mark where she was.

"We looked at the way those pieces of moss had been patchworked together, as if by hand. [We] just looked at each other for a second, and I could tell by the way my hair was going up on the back of my neck that something had gone on here in the night and I got a real eerie feeling. Mark looked about like I felt. Both of us felt that this cache had been done by something other than a bear or

Figure 40: Author's impression of black bear, reported shot near Upper Wolf Lake, Revilla' Island, AK, by witnesses the previous night and found early the next morning. She had been placed in a hole and camouflaged with large patches of moss.

another hunter, something strong, but with hands. We got out of there and had lots of other good spots for deer so just haven't been back." [B98B-map5]

Bears, and cougars where they occur on the mainland, usually do not conceal the entire carcass, but this is occasionally documented for brown bears. Brown bears do not, however, occur with any regularity on Revilla' Island. At any rate, the bear carcass was not touched or partly eaten in any way that the men noticed, and this argues strongly against a bear performing the action, although grizzlies have been known to cache whole black bear carcasses. That sasquatches can be inordinately stealthy around man is documented in dozens of reports from British Columbia, and contrasts with the usual behavior of bears when appropriating kills and carrying them off. Most people familiar with bears will tell you they make enough of a racket taking away a kill or getting into camp food. It is difficult to get a deer out of an area quietly while dragging it; only by holding it off the ground is it possible to move it without making at least some noise. It is hard to picture a bear quietly absconding with a buck in the manner of a retriever carrying a duck, and it seems rather more likely that something else was responsible. Even another hunter attempting the feat is likely to make a modicum of noise, unless, of course, he is wearing large cushioned pads on his feet to muffle the sound. At any rate, it would seem that sasquatches are fairly interested in deer, dead or otherwise.

On the coast, other food species tentatively associated with sasquatch reports include ducks, as noted by Bindernagel, and shellfish as stated in some Native reports. It is suggested by several reports that black bears may also be prey for sasquatches.

There is also one curious report from a Ketchikan man who was with a friend's family in the Misty Fjords area in the 1980s. The man reported to me that the entire family caught a glimpse of a brown figure in the bushes near them, followed amazingly by a regular-sized black bear, which they said they saw thrown bodily out of the thicket, with no roars or noise involved. The bear got up in a hurry and took off, they said, and they admitted they left the scene in a similar fashion. He concluded by saying that the family had no wish to discuss it further, but felt certain they were sincere and clear about the facts. Was a sasquatch responsible? It seems possible.

Thomas J. Fisher's 1985 Yakobi Island report of a large brown bear apparently fleeing from the direction of a large sasquatch seen minutes later, and J. B.'s 1998 report of a black bear carcass mysteriously cached and camouflaged overnight on Revilla' Island, lend credence to the notion that interaction does occur between the species and that sasquatches, adults at least, might have no hesitation in confronting black bears, while avoiding Alaskan browns.

Hypothetically, sasquatches and coastal brown bears must have some sort of mutual tolerance, or perhaps avoidance behavior, if they are each to avoid injury. This would certainly have to be the case for young sasquatches and perhaps adolescents, but might not necessarily hold for larger male sasquatches. In the Yakobi Island report there is the implication that for some browns at least, discretion is used in avoiding mutual confrontation. Perhaps sasquatches inhabit lower elevations when the bears are higher up in the spring, there being little opportunity for contact during bear hibernation during the winter. It may be that brown/grizzly bears and sasquatches have simply developed a variety of ways of avoiding each other.

In closing the subject of sasquatches and their possible diet however, I will recount a statement made to me by a Native Lummi woman in the old tribal band offices on her reservation near Nooksack, Washington, in October, 1975. I was there to investigate police reports of sasquatches, and had just followed Rene Dahinden, well-known sasquatch hunter, and tribal police into the building. In the rather narrow hallway, a diminutive but buxom tribal elder was squeezing past the two men in front of me, a frown on her face. Trying to navigate past me and my pack of recording, video and track-casting equipment in the hall, she confronted me closely for a few tight moments, unable to get past easily in the narrow passage. In exasperation, she reached up and shook me by the lapels, spilling my cameras, and demanded loudly, "Why are you always trying to kill these creatures? Why can't you just let them eat their mice and berries and be left alone?!" Her accusation left me speechless, but I have since regretted not telling her my interest was more in what they eat and less in the guns she witnessed.

What we may come to know of sasquatch anatomy will depend on eventual samples of bone material, blood or other body tissues. On the other hand, greater knowledge of sasquatch behavior will depend entirely on reported sightings of sasquatches, their tracks and signs. Anatomists may have to wait for their specimen, but behavioral scientists are under no such constraints. They must draw on a documented historical database—and the more detailed and confirmed reports the better. Although an anatomical specimen is the formal event giving any "undiscovered" animal its taxonomic name and legal status, it won't answer any of the hundreds of questions that must follow. This will require further analysis of sightings and also the examination of purported sasquatch nests, vocalizations, spoor and evidence of their activities. In the following chapters, we will take a closer look at more circumstantial evidence and how it may possibly be attributed to sasquatch behavior.

1. Bindernagel, pp. 116–120.
2. Bindernagel, pp. 112–114.
3. Hewkin, James, 1998, "Hanging Kills," supplement to *The Track Record*, Western Bigfoot Society newsletter, ed. Ray Crowe, Hillsboro, Oregon.
4. Green, John, 1987, *Sasquatch: the Apes Among Us*, pp. 333–336.
5. Bindernagel, pp. 107–131.

14 Possible Sasquatch Nests and Other Signs

"The nest appeared fresh, made that spring. But there was something more peculiar about the nest. What really caught my eye up above it, constructed into a sort of 'wickiup'—was a canopy made of the same sort of branches. In seventy years in the bush I've never seen a bear day bed with a roof to it! Whatever made that nest, it sure wasn't a bear!" Mr. Bruce Johnstone Sr., Ward Lake, 1960

Wildlife biologist Dr. John Bindernagel has listed nest building as a fairly common sasquatch attribute.[1] While there are few if any reported sightings of sasquatches being seen actually resting in a nest structure, complex nests made of interlaced branches and bark strips are periodically noted in reports from areas up and down the Pacific Coast. It is likely so large that a mammal would be able to forage natural resources in an opportunistic manner, staying temporarily in any one area that offered concentrated food supplies such as fish, tuberous plants, wild fruit, seafood or abundant game. Staying in one place for even a short period of time, whether for rearing offspring, exploiting concentrated food sources or both, would arguably increase the likelihood of nest-building behavior, if for no other reason than comfort.

Nest-building has been noted as occurring fairly often in mountain gorillas.[2] If a gorilla is able to perform the rudimentary weaving involved in sequentially bending down saplings, then an ape of equal or greater intelligence could easily be expected to be capable of nest making, using branches, mosses, ferns or grasses, as the available resources dictate.

The earliest report of what may possibly have been a sasquatch nest in southeast Alaska comes from Bruce Johnstone Sr., a man who has been mentioned in many local hunting, logging and prospecting stories. He is very familiar with all the wildlife to be found in the area's mountains, forests and waters and is even known to have survived being mauled by a brown bear. Although he does not claim to have ever seen a sasquatch, he states he has been puzzled more than once in the bush by sounds and signs that he cannot readily ascribe to any of the "known" mammals or birds of the Pacific Northwest. In the summer of 1960, Mr. Johnstone had taken a short hike with his daughter Sharon up a trail on Slide Ridge, west of Last Chance Campground. The location was just one mile north of Ward Lake, an old-growth recreation area five miles northwest of Ketchikan.

"In the bush at the end of a short spur was a game trail that led off up Slide Ridge and we followed that a ways. The bush was thick, the leaves were all out and it was dense vegetation in there—alder and hemlock. After a short ways we suddenly came on a nest about six feet across made of branches and such. There was a strong animal smell there and there was a pile of droppings right beside it. They could have been bear droppings, and my first reaction was 'Bear, we'd best turn round and get out of here.' You could see around the nest where branches had been broken to make it. There was a lining of leaves to the nest, no grass or plants as one might expect with a bear. The nest appeared fresh, made that spring. But there was something more peculiar about the nest. What really caught my eye up above it, constructed into a sort of 'wickiup,'—was a canopy made of the same sort of branches. In seventy years in the bush I've never seen a bear day bed with a roof to it! Whatever made that nest, it sure wasn't a bear!" [N60A-map5]

Mr. Johnstone and his daughter left and returned to their vehicle in short order. Curiously, this type of canopied nest has been previously reported as possibly associated with sasquatches by a grizzly bear researcher working on the central British Columbia coast near Knight Inlet.[3]

• 1972: In December, 1999, Mr. M. J. Hert, a Ketchikan hunter, reported a fresh nest he found in late August, 1972, which he did not think belonged to a bear. Curiously, he reported its location as less than a mile from the site of the structure Bruce Johnstone reported seeing twelve years earlier.

"I had been deer hunting on Slide Ridge, Revilla Island, and was up about seven to eight hundred feet above the Connell Lake Road. I had walked up, finding no game sign and in thick hemlock and spruce had come across a nest that was about two feet high, circular, six feet inside diameter by eight feet outside. At first I thought it might have been manmade. But when I got up and looked at it closely, I saw it was made of branches that had been torn off and there wasn't a cut mark on them. The biggest branches were two and a half inches thick. There was no smell, no hair in the nest. It was weird, I couldn't imagine a bear making a big fresh nest like that in the middle of the year." [N72A-map5]

• 1984: Although there is no real evidence that sasquatches habitually use caves, a few nests in caves or abandoned mine shafts have previously been reported, sometimes in association with a sasquatch sighting or tracks, in British Columbia, Washington and Oregon.[4] One such nest was reported in Alaska by the late Mr. Harold Omacht of Ketchikan, then a retired mill worker and avid photographer. In an interview in 1996 he described something curious he had seen while spelunking the right-hand drift at Mahoney Mine, an abandoned mine site located twelve miles northeast of Ketchikan, where Mahoney Creek empties into George Inlet. He stated that in 1984 he had gone into the shaft to its end, a distance of perhaps 1,000 feet. Mr. Omacht attested that he found a nest at the end composed of boughs and branches in a fashion that seemed deliberately constructed. When asked about bears, he added, "No, the way the thing was put together would take hands to do." He added, "There was no sign of human habitation or use, nothing at all to indicate a human being had been there." [M84A-map5]

The shaft has since fallen into an unsafe condition and is now fouled by sour gas, precluding further investigation there. The nest may or may not have been fabricated by a sasquatch, but it is certainly not a particularly habitable spot for a human being to withdraw to, even a recluse. Nor are there any local stories referring to such a hermit, as one might expect in such a case.

• 1988: There is a remarkable report of a possible sasquatch nest discovered in 1988 near Klawock Lake, Prince of Wales Island, by a Ketchikan timber cruiser and logging engineer, Eric Muench. I investigated the circumstances of its discovery there and found that the report had initially been dismissed by his employers. When a logger first brought word of it to supervisory personnel at the local

Klawock logging office, curiosity about the nest aroused a number of forestry professionals and a state biologist to examine it. Although I have interviewed four of the men involved and their reports corroborated each other's, so far none of them wish to attract publicity and I have respectfully retained their names in confidence. Mr. Muench consented to interviews in 1998 and again in 2002. The following is his report:

"I had been on Prince of Wales Island working as an independent timber cruiser and logging engineer. On January 26, 1988, on a job for [a local Native corporation] on their land, I was on a hillside above Klawock Lake doing timber reconnaissance to plan some logging units for their coming season. It had been a fairly open winter, and there had been less than one foot of snow under the western hemlock and western red cedar forest at the five hundred foot elevation.

"I noticed a patch of huckleberry bushes on the hillside below me that had been broken off uniformly at the four or five foot height. Looking closer, I found a large nest of crudely woven huckleberry branches and cedar bark strips and boughs, lined with mosses and more bark. The circular nest was about seven and one half feet on the outside with a four and one half foot diameter hollow part inside. It was uncovered, but well-placed on the lower side of a downhill leaning red cedar with lots of live feathery boughs hanging directly over the nest, like a natural shingle roof. It was on about a ten-foot wide gentle bench, beyond which a series of small cliffs dropped on down the hill. Nail or claw marks on the tree showed where materials has been gathered, and the surrounding ground was stripped of grasses also. The site was less than one-quarter mile above the Klawock Hollis Highway.

"In my experience," Muench continued, "most bears hibernate in a convenient windfall den, hollow tree or similar partial shelter, with little or no preparation or housekeeping. I have also seen where mother bears will pull in moss, grass or brush tips, probably to warm and soften the place a bit for their cubs. However, this was quite different. Not only were the nest materials somewhat woven together in a way that no bear could do, but the huckleberry bushes had been broken off cleanly, as though two hands had bent the stems so sharp that they could not splinter.

"I wandered around the area a while to look for tracks on deer trails and passages through the cliffs, but the snow was mostly fresh from that day and still falling, so I found nothing. I did pull some fairly stiff, long and slightly kinky black hair from the nest and saw what appeared to be a louse egg on one. It reminded me of horse mane hair, not bear or wolf. The scratch marks, to about six or seven feet up from the ground, clearly showed individual handpulls. The scratch spread was about eight inches, similar to my own fingers if I spread them way out, but at that spread I could not put scratch-making pressure on my thumb and little finger. I tried, and could not begin to match those marks.

"While continuing logging road location work the next day, I visited the site again. It had not been disturbed. I designed the logging layout so that the immediate area of the nest was included in a timbered leave strip that protected a deep gorge nearby.

"Because I had recently read a down-south 'bigfoot searcher' declare that he intended to prove their existence by offering a reward for a shot specimen, I was reluctant to spread word of my find and risk 'outside' [non-Alaskan] clowns crawling all over my client's property. However, I decided that two people had a right to know...[Mr. Muench named a former land manager for the Native corporation and the logging superintendent for the privately contracted logging company.] I knew them both to be honest, intelligent and thoughtful men, and had no hesitation in letting them decide how far to spread word of the nest. Both took the news calmly and without skepticism. In the following days I heard accounts of frequent past sasquatch encounters, including both the Tlingit and Haida names for them, mostly from Native people who had grown up in the Craig and Klawock area. Apparently, knowledge of and belief in bigfoot is common in the area, but not often spoken of to strangers from outside the area.

"On February 9, during a Forest Practice Act inspection, the land manager and I took an Alaska Division of Forestry forester and an Alaska Fish and Game habitat biologist to the nest. The biologist gathered a sample from some unfamiliar (at least to me) small dropping piles.

"Later that spring or summer, I returned with a camera to photograph the nest and scratch marks, etc., using my ax and a six-inch ruler for relative scale on the pictures. By that time the nearby brush had 'leafed out' and the boughs in the nest, originally green, had turned brown.

"My only other observation of anything unusual in the area was

Figures 41 and 42: Photographs of eight-foot nest reported by loggers east of Klawock Lake, Prince of Wales Island in 1987, ax for scale. *Photo: Eric Muench*

that, on several occasions during that time on that hillside, I heard a series of slow, measured raps, as though a heavy wood chunk was being swung against a tree. I work alone, and knew that there was no other person anywhere near the area. Following in the apparent direction of the sounds never revealed anything. Years later I was told that such rapping has often been associated with bigfoot sightings or evidence."

Mr. Muench concluded: "I am aware that other people in various capacities visited the nest afterward and before it became destroyed by 'wind-throw' and fire. However, none of that was part of my experience."

Figure 43: Photograph of large cedar beside Klawock Lake nest showing bark stripped twelve feet up tree.
Photo: Eric Muench

Dick Hamlin, a Ketchikan pilot and retired teacher, maintains an interest in the research involving the nest. He has retained two samples from the nest, a dark tapering hair about four inches long and a plastic container with frozen fecal material. The hair has been studied and video photographed at 400-power magnification in Ketchikan, but neither sample has been sent out of state for DNA testing. DNA studies might certainly prove useful if someone with funding becomes interested, but are quite expensive, up to $5,000. One of the logging supervisors mentioned that the had accompanied several state biologists to the nest and that they had conducted analysis of some of the hair, in which a species of louse was verified as known only to be associated with a species found in Asia. Two of the men said that almost all the hairs had been studied at a state lab and when the samples ended up not matching anything, the lab people were alleged to have said: "We threw it all out." At least three of the men who discussed the nest with me were also interviewed by Al Jackson in the late 1990s, and he made the comment to me in 1999 that in at least two other photographs which he had seen (not those of Muench), the huckleberry and small strips of cedar in the nest appeared not simply woven, but "actually braided together." According to Jackson, one

Figure 44: Parallel eight-inch span marks in cedar at Klawock Lake next where bark had been stripped. [N87A-map4] *Photo: Eric Muench*

Ketchikan biologist, who showed much interest in the nest, said it resembled "gorilla nests" that he had seen in Africa. (Photographs taken of the nest, taken at the time of its discovery appear in Figures 28–30.) [N88A-map4]

The nest has long since been destroyed by a local fire, wind and by the large cedar which fell over it. Curiously, the loud wooden rapping sounds Muench reported in the same vicinity around that time have been reported on both Revilla' and Prince of Wales Island, in fact as far south as California. But coastal reports of this type of nest are not unique in Alaska or British Columbia. The circumstances surrounding this nest, and its similarity to others of its type in the Pacific Northwest, seem to suggest it was almost certainly fabricated by a sasquatch.

• 1988: Mr. Jim L., another Ketchikan hunter, stated in an interview in 1999, that he had found an interesting nest in early fall 1988, on Revilla' Island. It was, he said, on the divide separating Harriet Hunt Lake from Leask Lake (less than ten miles north of the previously reported nests).

"It was circular, freshly made of cedar boughs, huckleberry branches, was woven in appearance, and was only about four feet in diameter. It seemed a few months early for a bear to be making a big nest, and didn't look to be in a very good spot for hibernating." In it, he said, he found black hairs, "longer than a bear's, well over four inches in length." What kept puzzling him more, he added, was that "it was not from the previous winter, but freshly made." [N88A-map5]

- 1991: Reports of puzzling nests found over the west half of Revilla' Island continued into the nineties. In 1994, M. A., Ketchikan, a professional military man and hunter, reported seeing one such nest in 1991. I found him to be concise and consistent. When reinterviewed in 1998, he did not vary in any details.

"On November 4th, 1991, I had taken my deer rifle on a hike up Second Waterfall Creek, up off the highway north of Ketchikan, to scout the country and hopefully get a deer. I entered the thick alders along a faint deer trail west of the creek and had worked my way up to a saddle in big hemlock and some Sitka spruce. This was southwest of the creek and at an elevation of about 1,000 to 1,200 feet. I had taken a ninety-degree turn off the deer trail, which paralleled the creek, to gain elevation and look for deer sign. I didn't notice much sign so had turned back toward the deer trail which was down slope about 350 meters [1,100 feet] below me. I was in fairly tall timber and the undergrowth was not too thick, so I thought I would make a beeline straight downhill, jumping a bit here and there to save time.

"I had just jumped down onto a large collection of brush lying tight in against the slope when I was surprised to find myself in what I can only describe as a huge nest. It was about twelve feet in diameter on the outside and was put together with shredded bark and logs and branches. It was really deep and my head and shoulders barely stuck out over the top, which was a thick outer rim of heavy branches laid down in a solid semicircle. Some of the wood was three inches in diameter and about three feet long. It was like a big bowl-shaped affair that had been constructed by something strong enough to handle the hefty logs, but careful enough to intertwine the rest. There was no smell or feathers or hair, for that matter. It wasn't so old as to have any moss on it. It wasn't wet or uncomfortable at all. I noticed I had a really great field of view from there. Nothing could get at me easily from below without making a real racket and it was pretty sheltered. Sitting up in there, it occurred to me that it would be a good location to spot game below. When I climbed down out of it though, I saw that it went down slope about fifteen feet from the lip; it was really huge and I figured that it would not be anything a hunter would ever bother making." [N91A-map5]

It may be worth noting that bear day beds are not this elaborate, and I do not know of any bear biologists who have documented four-inch logs in the construction of bear winter hibernacula, although this is perhaps remotely possible. Furthermore, bears most often prepare a winter den under a root wad or log or dig into the ground, and the site is chosen where snow will accumulate above and provide some measure of insulation. Nests attributable to sasquatches, on the other hand, seem to have nothing to do with good hibernation locations.

- 1992: There is one second-hand report from B.C. obtained in 1999 through a Ketchikan woman, Kelly Needham, who mentioned a similar nest reported to her in the mid-nineties while visiting relatives near Massett, Queen Charlotte Islands. She stated that, while visiting Massett, she was told by a Haida relative there of a nest found in late summer or early fall of 1992. Kelly quoted the woman, a well-known and respected teacher of Haida crafts: "It was a very large, peculiar, ten-foot diameter circular nest, not a fallen eagle's nest, built in the forest near Agate Beach east of Massett." [N92A-map7] While the report is vague, and there is no way of confirming it, the nest does match descriptions of others found in areas of sasquatch reports.

All of the great apes, except gibbons, indulge in nest-building behavior to varying degrees. That gorillas build nests on the ground in the volcano rain forests of Rwanda-Urundi, eastern Congo and western Uganda has been well documented by George Schaller, a well-known field biologist, ecologist and author. He noted that while both male and female mountain gorillas construct nests for sleeping, the males more often build theirs on the ground. Their habit of bending down or breaking off large branches in a circle is something not attributed as a rule to bears, although perhaps this may happen on occasion. Likewise there seems to be, at least in gorilla nests, an initial focus on the circular outer rim, to such an extent that this is often the only part of the nest the gorillas may complete. Freshly constructed, high, circular nests found in bear population areas of North America built outside the time frame for prehibernation bear activity would suggest that something else is responsible for their creation.

The winter beds of bears, or hibernacula as they are referred to by biologists, pose the only possible confusion with sasquatch nests; they are usually under an overhanging log, root wad, ledge or embankment where the bear can pass the winter. In some places, both black and brown bears will line a hibernaculum with various materials, usually grasses, but these are often at higher altitudes where snow can insulate them. Hair samples from nests are suggestive, but not conclusive, since either bears or sasquatches might conceivably occupy the others' vacant nest for a short period. All such structures are subject to the ravages of time, which include local forest fires and falling trees.

Nests found on the ground that might be attributed to eagles are similar, but in the process of falling from a tree, almost always expose eagle feathers or down, even in old and weathered nests. It has often been noted that eagle nests that fall are badly broken up. Eagle nests are also usually lens-shaped and do not consistently show a cupped profile in cross section, as do typical coastal nests attributed by some to sasquatches.

The recurring reports of nests of a type similar to a gorilla's but larger, made freshly in spring or summer in the forests of the Pacific Northwest, are becoming more frequent. Where there is a difference in shape from a lenticular eagle nest and no likelihood of the nest having fallen undisturbed from a tree, eagles may be excluded as the original occupants. Similarly, bear hibernacula are most often excavated or partly so. Bears' dens and nests are also fairly often situated at an elevation and facing to provide a canopy of snow overhead for insulation purposes. Where these attributes do not apply and the nest structure appears to be made out of season for a bear preparing for hibernation, bears may be considered unlikely occupants. A construction using small logs in the foundation and a canopy may also be suggestive of fabrication by a sasquatch, if human fabrication is ruled out. One does not need to be a biologist to recognize that associated nest hair longer than a bear's may also be indicative of sasquatch activity. It is even possible that a good many nest structures previously attributed to bears or eagles may in fact belong to sasquatches. In future, an examination of such structures for hair, fecal material or nearby tracks would likely be useful in gathering information relating to population estimates for sasquatches.

Trees as Possible Sasquatch Signs

The relationship between sasquatches and trees is striking (no pun intended). That sasquatches snap trees or break them off as part of a threat display, displacement activity in anger or as a more pacific territorial marker has been mentioned by field researchers as possible explanations for this behavior. Trees twisted down and trees with large amounts of bark peeled off with no marks of claws or antlers are seen and mentioned regularly in association with sasquatch reports from Alaska, British Columbia, Washington and elsewhere.[5]

A most unusual arrangement of inverted trees, jammed thirty yards apart, into a muskeg in perfectly vertical fashion, was found above Klawock Lake, Prince of Wales Island, in the early nineties. As reported to me in 1996 by Klawock forest workers, the trees were located thirty feet off a logging spur, several miles up from the Klawock-Hollis Highway, which transects the island. According to researcher Al Jackson, Prince of Wales Native elders had stated that fifty years ago the trees had been jammed into the soft muskeg by huge two-legged creatures as markers. [N99A-map4]

The trees above Klawock Lake are all trunks set quite vertically, with root wads uppermost, in a seemingly deliberate fashion. They are what are called "blow-down" cedars, with much of their bark gone, and with accumulations of soil still remaining on top among the roots. The largest tree shows approximately thirteen feet of wood above ground; the next largest shows approximately nine feet. This more northerly tree still has soil throughout its roots nine feet above the muskeg. An estimated one-third of their total length may be embedded below in the muskeg. Visiting the trees in August, 1999, I detected no wobble to either of the large trees or any choker marks or grapple marks at all anywhere on the trunks.

According to researcher Jackson, they were first reported by his uncle, Mr. H. M., who told Jackson that he had "seen them up there before the logging roads went in there in the late 1980s."

"My uncle and his buddy Richmond Benson would go hunting and fishing with Old-Man Albert Brown of Klawock, who passed away back in the early sixties. Mr. Benson has also passed on. My uncle also told me that Old-Man Albert Brown had warned them that, 'whenever hunting above the lake on Klawock Mountain, you

Figure 45: Photograph taken in 1999 of Al Jackson and a twelve-foot cedar tree with root wad uppermost, inverted and thrust into muskeg east of Klawock Lake, Prince of Wales Island.

Figure 46: Photograph of nine-foot inverted cedar near larger upside-down tree, Klawock Lake.

have to watch out for those big black gorillas that live up there. They mark their territory by driving blown-down trees into the ground, upside down, with the root wads up in the air.' Of course, people thought he was making up stories, until they built the logging roads up there forty years later and found the trees."

On examining the trees, it certainly seemed possible. However, this is no proof of how they came to be inverted and thrust into the soft muskeg in a vertical position. No one I know who has seen these three amazing "upside-down" trees in the muskeg has suggested that they were placed there with conventional hydraulic logging equipment. There are simply no grapple or cable marks on the trees. Nor are there the telltale scars of hydraulic tread marks anywhere, something the muskeg retains for decades. Although helicopters occasionally drop trees by accident, logs from helicopter "dropped turns" are invariably set into the ground at an angle, due to the nature of their angled suspension by a strap coupled off to one side of the trunk. The reader would do well to examine the photographs in order to form his or her own impressions.

In addition to the two large cedars, there is a smaller tree, with roots uppermost, jammed into the ground right up to its root wad about ten feet south of the largest tree. According to a Klawock man, T. R., who described the trees as he first saw them in the early nineties, "The smallest tree was thirty yards north of its present location, in between the two larger trees, and only jammed in six feet up from the ground."

"Later," Jackson commented, "when I moved back to Klawock around 1998, the little tree [only seven inches in diameter] was jammed in where it is now, south of the other two, and right into the ground with just the root wad showing." There is certainly a hole in the muskeg between the two big trees. I have heard rumors of two helicopter loggers who, over beers, claimed to have performed the arrangement of the trees, and they were tickled at how the town of Klawock thought otherwise. But no explanations of such were ever let out. Professional helicopter contractors I have interviewed, showing them photographs of the trees, professed skepticism that a helicopter pilot and crew would ever be able to drop the trees so vertically and calculated the helicopter time and cost as something their budgets would certainly not permit. It may well have been possible, but I have not seen anyone perform similar feats.

The trees themselves are still standing. There is an understandable disinclination on the part of any company owning logging rights for the area to lose access to timber on the basis of what some might choose to call an "endangered species." This may also explain a fourth, obviously humanly cut, log rammed into the side of the logging road a quarter mile downhill, showing both hydraulic and saw marks. It is much smaller than the two larger "root wad" trees and obviously cleanly cut at the top, with no root system projecting at the top. It appears to be an attempt by someone to imitate the larger trees further up in the muskeg. As to why, I have no idea. This last obviously human artifact might be meant to imply that all of the trees are the work of loggers, so no need to investigate further. Or perhaps, this outlying log represents just a bit of idle play by some hydraulic operator to see if he could duplicate the big trees. So far, no vandals, loggers or pranksters have altered the posture of the root-wad trees, and they still stand with no marks on their smooth sides. They are immense and—in the case of the largest—probably weigh a thousand pounds. If they are a hoax, then they are equally remarkable for the skill involved, the absence of any marks or faint scars at all and the pristine, unmarked surface of the muskeg. Until some such helicopter logging team can step forward, the late Mr. Brown's original assertion, that they were placed there by huge hands is certainly as good as any simple conventional explanation. [N99A-map4]

Sasquatches and sasquatchlike creatures from Native ethnographies have long been credited with the habit of striking or knocking down trees, or even of twisting the tops to mark their territory. Marking a tree is certainly the easiest way for the human species of primate to reconnoiter in the backwoods, at least until the advent of Global Positioning Satellites. The following Alaska reports are some examples of activity that would be most easily attributed to sasquatches.

In southeast Alaska, an early reference to this alleged behavior came to me from Bruce Johnstone Sr. who stated the following in an interview in 1996.

"At times, in the thirties and forties, up the Unuk and Chickamin Rivers, I have seen trees twisted off, four inches or more in diameter, that showed no signs of man, bear or moose having done so." When I questioned him further about the possibility of natural phenomena such as windfall, snow load, avalanche or rock

Figure 47: Photograph of green, unmarked alder bent and snapped across Carlanna Lake Trail, August, 2001.

fall accounting for these, he remarked, "Not in any way that I have ever seen." I was assured that he was quite familiar with these forces of nature, having seen them all before for himself in numerous localities in southeast Alaska, and he was quietly adamant that these were not the case. [NS30-map3]

That all the large great apes habitually bend, break off or twist branches, saplings or small trees has been widely noted, often in association with aggressive displays and threat behavior. Swinging such objects against tree trunks has been mentioned in field studies on chimpanzees. Trees that have been altered or uprooted and repositioned are documented in the Pacific Northwest but are more difficult to attribute solely to sasquatches.

Identifying broken trees as the work of sasquatches may be misleading. The problem is apparent when one examines the work of ungulates such as moose or elk where a sapling has been twisted off for the purposes of shedding velvet from the antlers and marking territory or dominance. Moose, which occur on the mainland of Alaska but not the islands of southeast Alaska, often break off alder or aspen for browse by straddling them. Browse marks made by such animals are evident on the tips of the newly accessible branches. Twisted-off trees regularly attributed to sasquatches show no such signs of feeding.

In early August, 2001, I accompanied my seven-year-old daughter on a hike to Carlanna Lake, above the northwest corner of Ketchikan. It was a bright summer's day, and we had the trail to ourselves. As soon as we entered the forest we were stopped by a freshly fallen green alder, four inches thick and forty feet in length, that lay snapped across the trail. Surmising that a group of boys had climbed it to bend it over, I looked for any bruise to the bark that would indicate it had been climbed by man or even bear. There were none. The tree had been snapped off about four feet above the ground within days, and the recent weather had not been windy. There were no signs of feeding or browsing.

The next day, retired handlogger Mr. Bruce Johnstone Sr. accompanied me to the tree and after an hour of examining the tree and surroundings, stated that he was puzzled. "It wasn't a bear that did this, and I don't think human either. I just don't know what could have done it." Curiously, a man I work with, a hunter, told me that late one evening in early August a half-mile up the trail, along Carlanna Lake, he and his family had been puzzled by repeated loud "splooshing" sounds that came from some distance around the corner of the lake. He didn't think it was Revilla' Island's only large known mammal, the black bear. Although Ketchikan hunter M. A. had reported a brief sasquatch sighting and a track one-half mile west in June, 1993, the tree and strange sounds at Carlanna seemed rather circumstantial. However in November, while reinterviewing Mr. Rod Stockli regarding his report of possible sasquatch vocalizations at Last Chance Campground in the late seventies, his wife mentioned that their teenage son and four friends had returned from an overnight camp at Carlanna Lake around July, rather disturbed about a sound they said had awakened them around 2:30 a.m. the night of their camp-out.

"The boys said they all woke up to hear what sounded to them like 'a baby crying loudly from the forest.' They are used to all the sounds of ravens and other animals around here, but it wasn't anything usual, and I think it really spooked them." Although the boys couldn't have been aware of it, vocalizations of a loud nature, much like a baby crying, have been attributed to gorillas by scientists and to sasquatches by many Natives. With such a variety of circumstantial evidence surrounding Carlanna Lake, it is quite possible that the forty-foot green alder bent over the Carlanna Trail in August was evidence of territorial or route-marking behavior by a sasquatch, if nothing else.

Signs such as trees broken, twisted or visibly altered in ways inconsistent with other North American mammals, especially in patterns, may also be significant artifacts of sasquatch behavior. Any wood material registering a large dental impression not displaying a gap between the lateral incisors and canines should be considered of possible hominid origin and forwarded to biologists or anthropologists actively pursuing research on sasquatches. The deformation of large trees associated with sasquatch sign, i.e., tracks, hair, etc., may be viewed as either territorial marking, locator display for mating or other purposes, idle play behavior or possibly part of a ritual dominance display. They may also be markers for types of food resources. Possibly they represent all of the above. Perhaps, like gorillas, sasquatches just smash vegetation when angered. Clearly intrepid and rigorous field work is required before much more can be said.

Rock Cairns and Other Possible Sasquatch Signs

There is one other phenomenon that may be attributable to sasquatches as easily as to man in that it requires at least a pair of hands. This is the stacking of rocks in a cairn or vertical pile. Bindernagel[6] and Green[7] have noted this behavior as sometimes observed with sasquatch feeding activities. Of course, near logging roads and where hunters may wish to get more arm exercise than surveyor's tape would allow, man has been known to mark the wilderness with piles of rocks. Prospectors traditionally did this to stake a claim. Occasionally hunters or scouts would mark the land with rock piles, but tree slashes, stakes, spray paint or bent branches have usually taken their place. There is still no reason why humans might not mark a burial place in this manner, but a cross or the like is usually associated to indicate that a human was interred on the spot. Piles of rocks in the bush may still be manmade, but those erected by surveyors, Natives or pioneers usually show some moss or weathering to indicate that they are not recent.

James Nanez of Ketchikan, whose Revilla' Island sighting is noted previously, added during his interview that around 1998,

Figure 48: Photograph of rock cairn discovered above Klawock Lake near the "upside-down trees," Prince of Wales Island, 1999.

while deer hunting with his wife on the north end of Prince of Wales Island west of Salmon Lake, they had noticed strange foot-high piles of rocks at intervals along the deer trail. And eventually they came upon a huge pile that looked too heavy for a man to have built. James said his wife noticed his hair standing up, something that puzzled the couple, James being used to seeing bears. Deciding to retrace their route back to their vehicle, James stated that whatever it was, the noises followed them on both sides, stopping when they did, resuming when they moved. They made it back to their truck without catching sight of anything. [N98A-map3]

In August, 1999, while returning from a trip up Klawock Mountain near the site of the inverted trees on Prince of Wales Island, I stopped to examine three piles of rocks that had been neatly constructed on the side of a gravel logging road. They were composed of twenty-pound and smaller rocks, not easily discernible to the human eye from a passing vehicle, and not likely the work of on-the-job loggers, although this is certainly possible. The only other likely candidates might have been hunters who wished to covertly mark a spot for leaving game to cool (not wise in such prime black bear habitat) or to mark where one might park and head off through the adjacent clear cut. However, it was certainly not the best terrain to head off through! As well, if an early morning hunter (or one arriving late at night) were to use these as markers, he would have a miserable time trying to spot them in the dark. Not that they couldn't easily have been made by humans, simply that it's much faster to either tie off surveyor's tape to a tree or shrub or else to break down one or more branch on a sapling. Researcher Al Jackson, in returning to the site weeks later only to find them gone, speculates that the cairns may have been placed there by passing male sasquatches as markers for others to follow and were later removed by the females and juveniles coming behind them. Speculation admittedly, but an interesting hypothesis nonetheless. [N98B-map3]

Retired Washington law enforcement officer, the late Fred Bradshaw, who had noted the presence of such cairns in areas of alleged sasquatch activity, had even suggested that such cairns may not be markers at all. He believed they could be thermal traps set up to radiate heat in the evening, attracting nocturnally active mammals, such as voles, to be gathered there by a returning sasquatch in the night or early morning. As far-fetched as this may sound, it is no more unreasonable than saying cairns are calling cards for sasquatches or a territorial marker of some sort. In any event, the Alaskan cairns were found within a few miles of reported sightings, a nest and "log-rapping."

Several outdoorsmen from Klawock have stated that they have seen a number of such cairns, associating the cairns with the tracks and occasional sightings of large male sasquatches, whom they also attribute with marking their way for females and juveniles of their family units to follow. The men, employees of different local Native corporations, did not wish their names disclosed, but were all of the opinion that the cairns were disrupted or taken down by the females as they followed the male. This, too, may sound far-fetched, but the pygmy chimpanzee has been alleged by researchers to use trail locator markers, and it may be that other primates besides ourselves do so on occasion.

Ketchikan prospector Jake Lauth, noted to me in the fall of 1998 that on the east side of Thorne Arm, southern Revilla' Island, at a place on the map called Elf Point, he had previously noted the appearance of a three-foot rock cairn at the edge of the forest above the beach. He has anchored just a few miles north of the spot regularly for years, having the only mineral claims thereabouts, and is always aware if anyone has come in. He told me that he did not think the cairn was likely manmade, and has had other experiences in the area that he cannot easily ascribe to bears or man. Cairns of rocks are intriguing, but unless accompanied by tracks, hair or manually dismembered game, they are difficult to ascribe to sasquatches rather than man. Although they are represented in a small but significant number of reports, their purpose remains poorly understood and some of what has been said regarding "sasquatch trees" may also apply to cairns. In the meantime, they are as attributable to sasquatches as to anyone, as well as being just plain interesting.

Field reports of alleged sasquatch behavior by any researchers trained in anthropology or biology would certainly not be intended to satisfy colleagues more interested in anatomy or genetics. It is without doubt the most useful tool at this time in helping us gain an understanding of "what is it that is out there." Each branch of biology has its own criteria for determining what is acceptable sasquatch evidence. Tracks, film analysis and fingerprinting have all produced strong evidence of sasquatches' existence beyond old folklore, but few academics can be faulted for not wishing to hang all their coats and hats on one hook.

Conservative anthropologists and zoologists would appear to be waiting for someone else to make the risky pronouncements, and while some wait for their own personal sasquatch to be delivered on their doorsteps, reports are accumulating in the thousands. By currently reviewing documented reports of sightings and also other sasquatch phenomena such as tracks, hair, alleged tree-striking, nest or cairn building, etc., we are gradually evidencing a profile of hypothetical sasquatch behavior that may prove useful as a baseline for further studies by field zoologists.

1. Bindernagel, p. 69.
2. Schaller, *The Mountain Gorilla*, pp. 171–199.
3. Bindernagel, p. 69.
4. Bindernagel, p. 70.
5. Bindernagel, pp. 73–76.
6. Bindernagel, p. 71.
7. Green, John, 1973, *The Sasquatch File*, p. 302.

15 Traces of Hair and "Missing Bones"

A collection of matching hairs found in association with alleged sasquatch tracks or sightings from more than ten different locations in western North America has been well-studied and described by Dr. W. H. Fahrenbach of Beaverton, Oregon. The result of Dr. Fahrenbach's work by itself is suggestive of a new species or subtype even without recourse to mitochondrial DNA analysis. DNA analysis is straightforward (and expensive) to perform on freshly obtained hair and next to impossible on old hair that does not have a hollow center or medulla (such as human or alleged sasquatch hair). It is quite impossible to perform on old samples that do not have a root or follicle—the growing, living "bulb end" of the hair that contains the only few real cells in the hair itself. It is this bulb, collected fresh and handled correctly, that is required for a DNA "fingerprint." And hair does not usually detach from the body complete with a follicle, unless someone has walked up to the living hair and plucked it or found it on barbed wire. Alaska, unfortunately, does not have much barbed wire.

All mammals have hair. And all known mammals may be identified, at least to the genus level, by the length, thickness, cross-sectional shape, medullary canal configuration and external scale pattern of their hair. Reports of strange hairs of unusual length or color, which by their description do not match that of previously identified mammals, are not common. However, Bindernagel[1] has noted that, "Hair attributed to sasquatches has been recovered from apparent sasquatch beds, shrubbery, fences and trees, all where sightings have occurred." It is some consolation that in at least a few cases some samples of hair have been kept. The most frustrating aspect of hair analysis, at least to sasquatch researchers, is the bland statement heard more than once from law enforcement and fish and wildlife departments who have analyzed the hair given them: "It didn't match anything we have on record. We threw it all out." In spite of this, there exists among many casual sasquatch researchers the naive belief that the same laboratory staff who couldn't match the samples would persist in the endeavor to find some higher agency who had a more complete collection of hair.

Two points arise here. Firstly, if more paperwork is necessary and the analysis of the hair isn't an official department task, the work ends immediately along with the sample. These good government (or, too often, university and museum) employees are not particularly interested in spending much of their time discovering a new species and work daily within the realm of what is already known to science, no more. Secondly, there are just no known samples of sasquatch hair with which to compare reported hair in any local fish and wildlife agency offices.

In southeast Alaska and elsewhere, there have been hairs regularly collected in the bush, on trees or in peculiar nests that do not match the long hair of bears, wolves, moose, porcupines, elk or wolverines. From Alaska, I have only three reports of possible sasquatch hair, including one with a sample currently under analysis. The mentioned collection of hair attributed to sasquatches has been subjected to microscopic scrutiny by Dr. Fahrenbach whose work has included specialization in the field of primate studies. While his collection contains more than ten hair samples of a peculiar and matching structure collected in association with reported sasquatch sightings, tracks and so on, he and his colleagues have found DNA sequencing to be frustrating and, at this time, perhaps not terribly useful. The recent state of matters regarding sasquatch hair may be best summed up in his own words in a reply to individuals seeking information about field collection of purported sasquatch hair.

Fahrenbach stated in February, 2000: "There is no absolutely identified BF [read sasquatch] hair. I have collected twelve samples from four states, all collected under suggestive or excellent circumstances. They measure from three to fifteen inches in length, are all under ninety microns in diameter, all have a reddish tinge under the microscope, and all are lacking a medulla. You can find human hair like that, but a dozen samples like that, collected independently, are not suggestive of hoaxing. DNA analysis is futile, since the mitochondrial DNA appears too fragmented to allow for sequencing [possibly a function of no medulla—same problem with human hair of that structure]."

In the most recent Alaska report of hair I have, the hair described was not too unordinary, nor was any sample collected. One Ketchikan hunter with more than twenty years of deer and bear hunting experience, stated to me in 1999 he had found some curious hair in a freshly made nest. The nest (described in the chapter on sasquatch nests) discovered in early fall, 1988, "was circular in form with a rough rim and although small, it was woven in appearance and was made of fresh huckleberry and cedar. I did not feel it was a bear nest. In it, I found black hair, longer than a bear's, well over four inches in length." Of course there is nothing to prove it belonged to a sasquatch, but given the poor time of the year for a bear to be readying a winter hibernaculum and the appearance of the long hair, it is rather more likely that it had been recently constructed by a sasquatch.

The second report of hair is similar. No sample was collected, but the details are quite a bit more remarkable. In a track report given in 1999 by retired Saxman fisherman Tom Abbott, then of Haines, the witness told me he had been seal-hunting with a friend in April, 1974, on Taiyasanka Inlet ten miles north of Haines. He further told me of finding an eighteen-inch sasquatch track on the beach by an area of trampled grass, blood and wolf tracks. What surprised him and his partner as much as the track was a big hank of light brown hair, approximately eighteen inches long. Mr. Abbott told me quite frankly he has since "really regretted not hunting with a camera or thinking to pick up some of the long hair." [T74A-map2]

Although shorter, four-inch hair is the usual length stated in sasquatch reports both in Alaska and along the B.C. coast. A number of both modern and historical reports up and down the coast report the observation of some individuals with long hair on the head, the back of the head and neck, sometimes the shoulders and occasionally down the back or the arms. Whether this unusually long hair represents a gender-based difference (male being the candidate on the basis of primate studies), a variation within the species or a different species entirely, has obviously not been determined at this time, nor will the acquisition of the first body or anatomical part solve that question. In the meantime it may be that some hair or bone will identify the genus and species of this hominid, and the next report will touch on that.

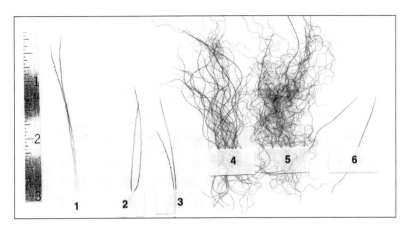

Figure 49: Photograph comparing hair samples from (left to right) moose, elk, black-tail deer, brown bear, black bear and possible sasquatch hair from Klawock Lake nest.

In a previously noted Klawock Lake nest report, a forestry worker discovered dark hair, nearby hand scratches on a tree as well as large droppings nearby. The nest was photographed, and the hair sent to a certain state lab was found not to match any of the known animals of Alaska and promptly discarded. The fecal sample, approximately one gallon in volume, was bagged and frozen, but has not been submitted for analysis and remains in the possession of one of the principal investigators. The nest was subsequently destroyed by a fire through the area and the large cedar fell on it as well. At this time nothing recognizable remains, except one single sample hair obtained from the Klawock Lake nest, currently under study.

The collection of hair in the field should be carried out with the possibility of securing a hair with the follicle at its base intact. This greatly increases the possibility of successful DNA examinations if the sample is fresh and forwarded promptly to a lab. Following are instructions for preserving the hair with a follicle. Carry a sealed bottle of sterile isotonic saline (0.9 percent saline), which is readily available. If more than a day from a lab, refrigeration would be useful to retard bacterial growth. The saline would function to keep the follicle(s) from dehydrating. If sent by air mail, the container should be no more than two-thirds full and taped closed. Shipping in a refrigerant, such as dry ice, should be adequate. Even if DNA cultures are not productive and evidence of the species is not obtained this way, the assembly of a collection of similar hair will probably prove useful in providing a reference for future possible sasquatch hair samples.

"Just Some Old Bones..."

The subject of bones always attracts a good deal of attention from even the most skeptical individuals. One often hears the skeptic comment, "No one has ever showed me any bones, so let's not hear any more about sasquatches." However, it must be noted that there are very few remains of any bear that a bullet did not bring down or that allowed itself to step in a trap. Any animal that seeks shelter when wounded, ill or dying, is probably going to seek terrain that is difficult in getting up to, into, over or under. Wolves work year-round to disperse, carry off and break up most of the skeletal material left over the winter in southeast Alaska and coastal B.C. On the mainland, porcupines and coyotes add to the cleanup. In spring, summer and fall, voles present a nocturnal army of appetites for the phosphorus and calcium contained in osseous material. About the only time one might find any remains of naturally dying mammals at all is in the early spring, prior to the increased activities of small rodents after the spring snows melt.

This being said, there are still reports, if not the samples themselves, of bone material that may have been related to a sasquatch. All three reports of bone material from southeast Alaska have been simply that, only reports, and I will be the first to say that as one with some training in anthropology, no conclusions may be drawn. For the sake of the record, I will here detail the three reports of such bone material, however anecdotal they may be, as they are certainly just as suggestive as many eyewitness sightings.

The first account was given by a professional Ketchikan woman interviewed in 1998. She recounted an incident remembered from her childhood in the 1950s on Annette Island, just south of Ketchikan.

"One day back on Annette Island when I was young, I told my father [the late Mr. John Milne] I had lost my new watch. He looked at me and said that made him remember the time he had lost a watch too, while deer hunting a few years previously. That would have been before the early fifties, and probably in the late forties sometime. He said that, some years previously, he had gone hunting by boat to a spot where he could get up high for deer, up some mountainside close to the shore, where he could just get the deer on his shoulders and slide down. I don't know where exactly, it might have been on nearby Gravina Island or Prince of Wales Island, or even on some part of Annette Island itself. Anyhow, he said that as he was sliding down the steep forested slope with his deer he got to the bottom of the slope and he had come to rest in the thick vegetation. He noticed his watch had slipped off and was just looking around where he stopped sliding for it. He said at that moment he got a real peculiar feeling, like the hair was standing up on the back of his neck, and he said that made him feel real odd. In fact, he said, he never used to get that kind of feeling and it was so odd he decided that after he got the deer back to the boat, just minutes away, he was going to come back just to see if he could find out the reason for it. He loaded the deer on his boat, took his rifle, got back to the spot at the base of the slope and started looking around for some sign.

"It was then he said that he came across something unusual. It was a huge skeleton, about eight or more feet long, and it was laid out inside a sort of low, rough arrangement of logs. The logs were just a couple of tiers high, laid on each other corral-style, no nails or rope or anything. The leg bones were long he said. When he held up his arms to show me how long the thigh bone had been, his hands were about three feet apart. He never went back as far as I ever knew and never told me anything more about them." [MB40-map5: location approximate]

Mr. Milne had died some years ago and I was not able to find anyone else besides his daughter who had heard his story. This account itself might not be remarkable except for the length of the thigh bones described and the curious structure described. The number of people over seven feet in height known to have roamed southeast Alaska is small, and a human could not easily have been the owner of a femur three feet or more in length. Acromegaly (or pituitary giantism) is the condition where an otherwise normal human grows in stature to seven or eight feet. It is known to occur in rare instances in Natives as well as non-Natives and may have been the cause of such a large skeleton. But Native burials were of a conventional sort in recent times and prior to that cremation was the rule, except for shamans who were usually buried in boxes. There is no mention of the logs being cut. And presumably a non-Native would also have a conventional burial; but in Alaska at the time, who can say? John Milne's curious encounter will probably never have a full explanation.

The second report is of a huge, roughly human-shaped jawbone reported examined by Bruce Johnstone Sr. and his brother Jack (since deceased) at Smeaton Bay (now in Misty Fjords National Monument) in 1948. This is the account of the jaw as told me by Mr. Johnstone in an interview in 1996.

"In 1947 and '48 my brother Jack and I were living in Smeaton Bay in a cabin at the mouth of Cabin Creek, shown on the maps as Bartholomew Creek. In 1948 we were hand logging a half hour's slow row east of the cabin in some big spruce. We had been working on one particular spruce, three to four feet in diameter located about fifty feet above sea level and about 200 feet back, and had stopped late one day for a break. About twenty feet from the tree, which is now probably just a big pile of moss on a rotted stump, I had stepped beneath a huge boulder, about ten feet by ten, which formed a flat roof over a ledge sloping toward the shore. There were smaller boulders, rock and dirt, forming walls to the structure. Under the overhang I noticed a jawbone, not belonging to a bear, but much larger than that of the average man. When I picked it up, I found it was so large it fitted over my entire lower face. [Mr. Johnstone was a large man; he stood six-foot-one.] It was not a fossil, in fact it appeared to have been exposed only several years and had bits of lichen growing on it. In the fading light of late day the growth made the jaw appear to give off a faint kind of glow. I think this may have upset Jack, and at his suggestion I put it back under the overhang, but further up at the back where there was a ledge that would protect it from the animals and the weather. Scattered down the slope among the moss and rocks we could see remains and fragments of the skull where they seemed to have slid down.

"It was getting late, the light was going, and Jack said 'Well, that's it, put it back, let's get back.' I think it may have bothered him a bit. So I just set the jaw up on the little ledge at the back of the overhang and we left.

"Now I've often wished I had gone back in there to look at it again. With the business of hand logging and such, I just never did. Some years back I told a Doctor Carr, then practicing medicine in Ketchikan, about it and he was really interested. He told me he would like to go have a look for himself, but I never heard if he did or not. A few years ago, I went up with friends, but my legs weren't quite up to the going up off the boat and so we couldn't tell for sure if it was the same place or not. I've told the past regional forest service archaeologist but I know they were always real busy and I don't believe they were ever able to get out there." [MB48-map3]

Mr. Johnstone's story is intriguing as he had never heard of any men that size in the country thereabouts and having lived, hunted, logged and prospected in the area all his life, he would be likely to know. He was clear about the jaw not being a bear and was only able to compare it to some form of giant humanlike being. He mentioned that the canines were not larger proportionately than a human's, and that the general appearance was one of a very large person. Without the jaw of course we are not able to make any assumptions, but it does lead one to speculate. Human giantism, or acromegaly as it is called, is one explanation. However, on three subsequent investigative trips by myself in the company of others, there was no sign of human artifact, no gun or ax to indicate human presence, and the two men were certain they were the first to log in the area. Natives have long used Smeaton Bay, but there were no signs of cremation or artifacts associated with burial, no remains of wood or any cedar box that might have been used to lodge the remains of a shaman.

To my knowledge, the jaw has never been recovered, although it was allegedly placed under an overhang. One of the forest service archaeologists then in Ketchikan, Mr. Frank Lively, said in 1997 he had been informed of the jaw by Johnstone but never had the time to get out there as he already had a large inventory of sites to visit. He has since retired and moved away. Of the medical doctor, a "Dr. Carr," working in Ketchikan in the 1950s, there is no current information.

Was the jaw in fact that of a sasquatch that had perhaps crawled under the overhang to die? One overhang there matching the general description certainly had all the appearances of just such a safe secluded spot. However, as I investigated the site several times in 1999, there was no sign of the jaw in or around it. Perhaps, as Mr. Johnstone surmises, it may be worthwhile to investigate down slope from that spot to search for dental remains, which would have conceivably passed the attention of voles, porcupines and other rodents fond of the calcium and phosphorus contained in bone remains. Perhaps archaeologists would like to try their hand at further field investigation at Smeaton Bay. The outcome remains to be seen.

The question of bones is often raised, but the bones themselves seldom are. If sasquatches are real, as many hold true, then why do we not hear more of any bones? Many people collect bone material of one type or another. Being collectors, most keep their collections quietly intact. For the skeletal remains of any undiscovered primate to become authenticated, bones or teeth must arrive in the hands of the nearest local anthropology professor or biologist with an interest. Interest and recognition are key here. Presuming the bone material gets that far it must then travel to a physical anthropologist or primate specialist, likewise with an interest, and be recognized for what it may represent. Police and wildlife technicians are not in the business of looking for, discovering or passing on any evidence of an "unknown" species. This is not to say that such material has never been mentioned or alleged to have been seen by people in the woods. It happens from time to time.

One such incident occurred recently near Ketchikan. Upper Ward Creek above Last Chance Campground near Ward Lake, the location of alleged sasquatch sightings in 1979 and 2001 and a possible vocalization report around 1978, was also the site of the following report. The account details the astonishing discovery of a skeleton of a small eight- or nine-inch humanlike foot. This report is really two separate accounts, both apparently describing the same collection of foot bones.

On November 7, 1999, a telephone report came in from Mr. Dale Morrison, a serious-minded Ketchikan sawmill supervisor. He described the remains of a foot he and a friend had allegedly found previously that fall. Mr. Morrison told me that sometime earlier that fall, he and a friend had been parked at the first bridge downstream from the dam at Connell Lake, where a small creek crosses under the road and flows into Connell Creek (also called Upper Ward Creek). They had been puttering below the bridge on the side of the feeder creek where ten feet of sand forms a flat bank to the feeder creek just yards from where it empties into upper Ward Creek. I have reinterviewed Mr. Morrison twice and found him straightforward and sincere.

"As we were going along the little creek, my friend and I saw a white cluster of bones in the sand and dirt on the bank. It looked, at first, like a human foot, just half-buried in the sand—right side and top showing with the ankle joint showing at the top. It was a right foot about seven inches long, and right away you could see by the shape that it wasn't a bear foot. It was about three inches across the ball and two inches wide at the heel. We picked it up. The first toe was the biggest, not like a bear's but it sure wasn't human. There was still skin and thickening, like pads on the bottom. There was also short black hair still showing between the toes, and in a few other places longer, one and a half to maybe two inches long. The pad color was gray. There were no claws. The four outer toes were all just bone but showed clearly, all in a row. The toes were weird. The big toe was flat and as wide as my thumb. We left it because my girlfriend wouldn't let me put it in the truck." [M99A-map5]

In December of 1999, a Ketchikan woman reported independently that she, her fiance and a friend had found a strange foot by the same creek below Connell Lake that same fall but had thrown it in the bushes on the bank. She said her fiance had intended to retrieve it later, but when he had returned it was not where he had left it. I spoke with the man by phone later.

"Around September we were at a small sandbar on Connell Creek with my girlfriend and friend and we found a foot that had a roughly human shape, but was about eight or nine inches long and had long black hair on it. We found it on a stream bank beside the gravel road just below the Connell Lake dam [describing the same exact location as Mr. Morrison had]. The foot had five toe bones and an attached ankle bone. It wasn't decomposed or smelling, but it had apparently been chewed on by animals. No sign of it being cut or shot, and the skin was missing from most of the top, showing the long tendons. It didn't look exactly like a young human foot or a bear foot either. There were no pads left on the toes, those only showed as the little bones closest to the joints on the ball of the foot, but the rest of the skin was grayish. It was wider than my wrist across the ball, about four and a half inches but a bit narrower at the heel, about two and a half inches, not pointed like a bear's. There was just a little hair left on the sides of the foot, black-colored hair. I started to bring it up to the truck but my fiancee said she would have nothing to do with that so I left it there. I just tossed it back into the bushes beside where it had been lying."

When asked to place the event in time he said he thought it had been late August or September, 1999. His girlfriend, interviewed separately, concurred, although she admitted she had not wanted to take a good look at it. Asked if he knew a Dale Morrison, her fiance replied that he didn't. When later asked why he had left it instead of bringing it back, he replied in much the same manner as Mr. Morrison had. He added he had been confident he would be able to find it by the bushes on the bank where he had left it when he returned, but admitted he had gone back up some time later and couldn't find it.

Though the little stretch of sand by the feeder creek was later investigated, there was nothing to be found by looking or digging there. There may have been a carcass or body buried there, but working the area over to a depth of six inches revealed nothing. I also spent approximately ten hours over several visits that fall, scouring Connell Creek in hip waders, but came up "empty-footed."

This leaves one with several thoughts. One is the apparent contradiction in dates. The second report involving the tossing of the foot into the bushes comes up first on the calendar or seems to. Neither party seemed exactly clear about the dates and this may be explanatory. Another problem is that there may have been two different remains. The fiance was not certain about whether the foot had been right or left. In both reports the reasons for not bringing it out the first time were curiously the same. This may seem rather coincidental, but it sounds conceivable. Of course, were one to attempt loading a dead foot into a family car, it's quite likely some passenger would find a similar objection. All of the people involved were relatively young, in their twenties, but on subsequent reinterviews, both parties stuck by their stories. There is of course the remote possibility that the foot was actually retrieved later by someone, perhaps without disclosure for fear of some legal involvement. In Alaska, it is considerably more probable that wolves, bears or ravens would account for its disappearance.

The reports agreed generally on the size, within an inch or so, and on shape, not that of a bear, but more manlike. Mr. Morrison's report especially emphasized the larger first toe, a human or ape characteristic, and the fiance's report had mentioned, "not a bear's, just like a human foot." They also concur on grayish skin color and black fur or hair, especially short hair remaining between the toes. The longer hair mentioned in the second report may have been related to a lesser state of decomposition at an earlier date, but scavengers may have accounted for some of the differences as well. Mr. Morrison's report mentioned sole pads of some sort while the second report stated "no padding like a bear's." The differences may be purely semantic. At this point we are far from certain knowledge of the nature of this alleged specimen, but the accounts do seem to refer to the same foot, apparently belonging to neither a human nor a bear. At this time there appears to be no forensic interest in these reports. And so, like so many other such stories, these two reports merely hint at what may have actually been a specimen of sasquatch anatomy.

Are there other reasons why sasquatch bones are not found—would sasquatches, like some bears, eat their dead? "Not likely," say most serious investigators. It is more likely, they say, that sasquatches simply do what many other animals do when injured or ill, simply retire to a quiet, secluded place such as under a large log, uprooted tree or rock crevice. Their remains would never be noticed. Some researchers have pondered the possibility of burial, and this at least is present in folklore from a few northwestern states.

Bones are, for many, the best evidence that science can use to confirm the species. Short of an entire skeleton, the skull, hands or feet would certainly convince the majority of anthropologists or zoologists. The vertebrae, pelvis and long bones, by virtue of their likely similarity to humans afflicted with acromegaly (pituitary giantism) would inevitably find themselves the proverbial "bones of contention." The second lower molars would be incredibly useful. It is these particular cheek teeth of man and the great apes that share a common pattern, and slight differences between the species are well known to anthropologists. These differences in the molars even enable scholars of early man to identify which of the many extinct species of man is being dealt with.

Eventually, in Alaska or elsewhere, some skeletal material will be uncovered and presented to serious researchers in a manner that will not allow such remains to be filed away as unidentifiable and, as such, conveniently disappear. At this time there are, to my knowledge, no samples of bone material available in Alaska which might prove to be those of a sasquatch. But this could easily change. It is certain that even a single, large lower second molar could establish sasquatches as living relatives of any known species of hominid. That would clearly give scientists, especially those too skeptical to bother with anecdotal evidence, something more substantial to chew on.

1. Bindernagel, p. 77

Part Three / What Could It Be?

16 North American Ape or Relic Hominid?

The average North American, if told today that a sasquatch had been found dead, sent to a university laboratory and declared by the attending professor to be "a species of ape," would probably not pause to question the "ape" label. Similarly, if the professor were to pronounce the sasquatch "hominid," or a species of man similar to *H. sapiens*, it would go down forever in the popular view as such. There is danger in either eventuality coming about prematurely. While scientists would be busy debating the issue, the general public will likely accept the first description given and never question with whom the sasquatch should be sent to reside. It is a puzzling question, and one that deserves a closer look.

Previously, Krantz[1] had included the following species, among others, as possible candidates for possible sasquatch ancestry: *Gigantopithecus*, *Homo erectus*, *Australopithecus robustus* (Paranthropus) and *Homo neanderthalensis* (Neanderthal).

There are three main hypotheses, which argue the following possibilities: (1) a separate species of upright pongid (ape) for which no fossils are known, or possibly a continuing species of *Gigantopithecus* migrated from Asia, Siberia perhaps, to the western hemisphere; (2) a larger, unknown form of a known fossil hominid, such as a larger form of *Homo erectus* (early man), *Australopithecus robustus* (Paranthropus, Zinjanthropus) or Neanderthal; (3) a completely unknown species intermediate to apes and man. Following are summaries of these three general hypotheses.

Sasquatch as an Upright Ape

Apes and sasquatches would appear to have a lot in common. Jane van Lawick-Goodall[2] has observed tool use among wild chimpanzees that

Figure 50: Author's illustration of possible facial appearance of *Gigantopithecus*, an Asian giant.

Figure 51: Author's illustration of facial reconstruction of *Zinjanthropus* (Nutcracker Man).

Figure 52: Author's illustration of facial reconstruction of *Homo erectus* (Java or Peking Man).

Figure 53: Author's illustration of facial reconstruction of Neanderthal.

includes "fishing" for termites with manufactured tools (twigs stripped of leaves). Primate researcher A. Suzuki noted that young Japanese macaques have been observed washing sweet potatoes, the older members of the group learning to do the same.[3] Chimpanzees have been seen using large sticks to break into termite mounds and tearing off branches in states of excitement. Male gorillas routinely break vegetation as a part of their threat displays. In 1983 British archaeologist and author Dr. Myra Shackley put forth her view that sasquatches were conceivably most closely related to the great apes, the gorilla in particular, and possibly descended from *Gigantopithecus*. Shackley, who gave serious consideration to such cryptic hominids as sasquatches, the Asian Yeti, as well as Asian cryptohominids *almas*, *gulib-yavan*, *mulen* and *chuchenaa*, suggested that the sasquatch may be less human than it is apelike: "In many ways the Sasquatch has diverged from its ape ancestors, and its behavior is certainly very different from that of probably its nearest living relative, the mountain gorilla."[4]

Sasquatches have been reportedly observed breaking off saplings and small trees as they move through thick vegetation. Orangutans have been observed dropping branches on intruders and are also noted to push over snags in similar situations. Gorillas are known to include throwing vegetation as part of their threat display. A. F. Dixson has summarized the entire ritualized gorilla display, some portions of which are performed by all ages and sexes of gorillas but is performed in its complete form only by dominant males.[5] These include, in sequence, hooting, symbolic feeding, rising bipedally, throwing vegetation, chest beating, leg kicking, running (bipedally) slapping or tearing vegetation and thumping the ground. While it is difficult to prove a sasquatch was responsible, there is one report from Prince of Wales Island of a log being thrown at people from out of the forest, and a number of rock-throwing incidents all attributed in suggestive circumstances to sasquatches.

Until recently, man was thought to be the only tool-using hominid. Gorillas and orangs have not often been reported using tools, but like chimpanzees, sasquatches have. Bonobos (pygmy chimpanzees) have been documented placing the long stems of vegetation at crossroads of their forest trails in West Africa, particularly in a manner which indicated to the observers that they marked the direction to be taken by following bonobos. Sasquatches have been allegedly observed pulling over saplings and branches in order to feed, but are also hypothetically attributed with twisting off saplings in a uniform direction, pushing over snags and leaving branches or rock piles lying noticeably on trails, purportedly as territorial or route-finding markers.

Other sasquatch behavior undocumented in apes, such as "vandalism" toward human shelters or artifacts, may be viewed as an extension of curious manipulation (as in orangs) or a display of anger (as in destruction of vegetation in chimpanzees or gorillas). Premeditated hunting of other primates or organized intraspecific warfare as described by Goodall[6] in Gombe chimpanzees is speculative at best in sasquatches, although there is some suggestion that sasquatches have followed humans, as indicated on several occasions by superimposed tracks. Chimpanzees, unlike gorillas and orangs, also kill and eat other mammals, displaying highly organized hunting groups.[7] Sasquatches have likewise been reported pursuing game, carrying game with a

Figure 54: Sketch of sasquatch reported carrying several ducks with a stick, on east side of Vancouver Island, circa 1969, drawn by witness for Dr. John Bindernagel. With permission, © Dr. John Bindernagel.

large stick in one hand,[8] digging on a beach with a bat-sized stick[9] but are not reported to be anywhere as group-oriented.

The chimpanzee's use of tools of stone, wood or plant material has been well documented. Hixson[10] refers to earlier research in the Ivory Coast of West Africa by Struhsaker and Hunkeler where wild chimpanzees were observed visiting particular "nut-smashing places." The apes placed the hard fruits of several types of plants in "depressions in the exposed roots of trees and employed sticks or rocks to break them open." Therefore, the occasionally reported use of such tools would not be sufficient to exclude sasquatches from an "ape" status.

Chimpanzees, orangs and gorillas have 48 chromosomes, we have 46, while gibbons have 44. The sasquatch chromosomal number is at present unknown. While the numbers may appear suggestive of the degree of relatedness between species, this is often not the case. Gibbons are very closely related to siamangs, but the gibbon has 44 chromosomes, while the siamang has 50. The numbers in themselves do not really prove anything, since in a relatively short span of evolution the numbers themselves may change through the normal process of chromosomal mutation.

The chromosomal "fingerprint " of a sasquatch would have to be taken from the body tissues or blood of a prior "type" specimen, one that was already known to science and available for further study, in short, the body of a sasquatch. When such a body is procured, the study of the chromosomal structure will be most useful in suggesting where to place the animal relative to the great apes and man.

Some fossil apes, such as *Gigantopithecus* (if it was actually an ape), are good contenders for relatives of the sasquatch. The first to clearly propose that *Gigantopithecus* is still living was John Green, in 1978.[11] *Gigantopithecus* was not only the largest hominid (ape or man) to have lived, but it somehow survived through at least two of the four last great glaciations. This huge hominid, known from only four fossil jaws and numerous teeth from western China and northern India, is arguably the longest surviving hominid in fossil history. *Gigantopithecus*, like the bipedal-quadrupedal *Sivapithecus* and the similar almost man-sized *Ramapithecus*, had jawbones that were not parallel-sided as in the apes nor arch-shaped as in man, but "flared,"— wider at the back than the front. While some anthropologists have downplayed the importance of this jaw flaring, Krantz has logically suggested it is related to uprightness. These differently shaped jaws with low, worn molars and thick humanlike tooth enamel, had given rise to the notion that they were ancestral to man. As summarized by Kay:[12] "Miocene *Ramapithecinae* (including *Ramapithecus, Sivapithecus* and *Gigantopithecus*) have long figured prominently as a group which may be in the direct line of human ancestry." This line of thought, originated by Weidenreich and von Koenigswald, was later discarded and for the past fifty years, anthropologists have sent the giant to reside with the pongids (ape side) of the hominid group.

Gigantopithecus is certainly sasquatchlike in its size. Its jaws are twice the size of a gorilla's yet with many characteristics that set it closer to man's early relatives. The projected stature of an adult male *Gigantopithecus* has been variously estimated at from eight to twelve feet, those of the early Dutch, Russian and Chinese providing the tallest estimates, while the British and American estimates are the shortest. Dr. Elwyn Simons,[13] a well-known American specialist on fossil man and primates, in 1968 estimated eight feet for height when upright, but suggested strongly that it was not erect but quadrupedal. Surprisingly, all of this hypothesizing went on in the absence of any skeletal material from below the mandible! It is rather amazing that few allowances for upright posture have been made.

Recently however, some authors have moved *Gigantopithecus* back closer to ourselves, including it, along with little *Sivapithecus* in its various forms, to "our" family, Hominidae, as opposed to the ape family or Pongidae. It is therefore now possible in some circles to refer to *Gigantopithecus* as a hominid (nonape) or as an early offshoot of that line, although some would disagree. Most everyone now places *Gigantopithecus* somewhere in between man and ape. Looking at the fossil jaws, it is not hard to see why. In spite of their huge size (one and a half to twice the mass of a mountain gorilla), the mandibles of *Gigantopithecus giganteus* have relatively much smaller canines than any of its relatives.

Those supporting the identity of the sasquatch as some new world form of *Gigantopithecus* has in the past included Green, Krantz and others. Scientific names, such as *Gigantopithecus canadensis* as proposed by Krantz on the basis of foot anatomy and gait biomechanics, have already been noted.[14]

Sasquatches as Relic Hominids

All fossil candidates for the sasquatch would require prehistoric migration to the western hemisphere during the ice ages or after, but only *Gigantopithecus* was known to be large enough at any time to match the reported size of sasquatches. The others, more manlike, would all have had to offer up individuals evolving to a larger size. Time to evolve is not really the problem. All would have had to migrate from Asia except *Australopithecus robustus* which would have had to migrate from Africa. Early migrating *Homo erectus*, or *Homo ergaster* as it is now known, has been recently shown to have migrated to Asia at an early date, perhaps one and one-half million years ago. Known from few fossils, it may have had a larger form that continued to migrate in the direction of North America.

There are a small but persistent number of reports describing the sasquatch facial anatomy as "human" or "humanlike." One particular drawing by Mr. James Edenshaw of Ketchikan (see figure 8), clearly portrays the sasquatch nose as convex as in *Homo erectus* (see figure 52), the first fossil man of our genus (*Homo*) to spread over the Old World. Apes usually do not have such convex nasal structures, nor did *Zinjanthropus*, a flat-faced relation of *Australopithecus robustus*. Close encounters with sasquatches are usually fleeting and it would be speculation to say how frequently, or infrequently, this shape of nasal anatomy occurs in sasquatches. John Green, who clearly proposes the ape resemblance, did not note any occurrence of this more "human" convex type, but close range reports in daylight are infrequent. Further reports may reveal more detail in this regard. As in our species, there may as well be room for a variety of sasquatch facial types.

Sasquatch teeth are also not usually reported as including large canines, as in the great apes. Furthermore, the Tsimshean mask collected by Emmons at the beginning of the last century (see figure 23) portrays the "sasquatch" canines as being quite even as in man, not pronounced as in bears, wolves or apes. In fact, the facial anatomy bears a rather striking resemblance to that of one form of fossil man, the Asian *Homo erectus pekinensis*, or Peking Man as he was sometimes called. This may ultimately be the explanation for some of the reports of more manlike hair-covered hominids in the new or old worlds, but it leaves us with a dilemma. Either there is one large species with a great range of variation in height, facial shape and weight, or there may be two similar large species coexisting beside man in North America and elsewhere, even possibly hybridizing. It is difficult to say.

Sasquatches, at the present time, don't qualify as any subtype of living *Homo sapiens*. They have been documented as playing with various human artifacts, occasionally carrying them off, but this is not likely anything more than curiosity. Use of fire is not reported. There exists one report where a sasquatch was believed to have played with the remnants of a dying campfire, flipping the last few burning pieces of wood about, but this is not recurrent in the literature and outside of a few scattered Native ethnographies, there is simply no evidence of their use of fire. It is the same with manufactured stone tools. But sasquatches, even as an early form of *Homo erectus*, would not have had any need for fire or tools along the resource-rich Pacific Coast. Indeed, humans in subsistence and survival modes have done quite well in the same environment on many occasions. In fact, it is plausible that a very early, albeit necessarily large form of *Homo erectus* or some other form of early man, even having discovered fire and simple tool using, might conceivably have lost these cultural adaptations in such a rich environment. The washing and stacking of foodstuffs occasionally reported for sasquatches, if corroborated, would suggest something more hominid than the great apes. These alleged attributes imply some awareness of the passage of time and future-oriented behavior, something not generally accredited to apes.

Longer head hair is a characteristic that is mentioned in a certain number of sasquatch reports. While not unknown in the apes, it is generally thought of as being a more "human" characteristic. As well, there are reports of longer hair on the shoulders, occasionally the back of the head and perhaps neck in a sort of "mane," and, in a few widely scattered reports, longer hair on the arms. Some folklore and modern reports match in this regard. The adult male orang, possibly the most closely related of the great apes to *Gigantopithecus*, has quite long hair on the arms. On the surface of it, this seems to suggest that there are some reported hair characteristics that are similar to man, and at the same time others which are characteristic of apes. Perhaps there may be several slightly different species of giant hominids in North America, one being a large humanlike creature, similar to but much larger than the Asian *Homo erectus*, the other a form of giant bipedal ape.

Gordon Strassenburgh of North Bend, a long-time wildman researcher, contended in 1974 that sasquatches are best represented by the African fossil hominid, *Paranthropus*, now considered a robust form of *Australopithecus*.[15] This upright and flat-faced hominid, known in one of its forms as *Zinjanthropus*, was man-sized but heavily built. Indeed in some fossils, its massive molars and jaws inspired the name "Nutcracker Man." Although, as Krantz has pointed out, there are the matters of size and transcontinental migration between any African fossil with the larger new-world sasquatch. Time allows for change however, and *Paranthropus*, by any name, has certainly had the time for any larger descendants to traverse Asia and arrive with early or late North American fauna. Although not highest on many earlier popular lists of "evolved" fossil candidates for the sasquatch, Krantz considered this species a possible candidate along with *Gigantopithecus*. As mentioned, he suggests that if sasquatches ultimately show cranial similarities to *Paranthropus*, they be given the taxonomic title, *Australopithecus canadensis*.

There are also several curious attributes that suggest something not typically apelike. One is the clear description of some nests that demonstrate not just simple weaving (although a few primate researchers have described gorilla nests in this fashion) but rather the construction of nests which had canopies woven overhead. Although all of the fossil candidates may have been capable of weaving a nest with a canopy, it argues a bit more strongly for the two of the most likely hominids, a hypothetically large form of *Paranthropus* or perhaps early *Homo erectus*, in other words, a species of large, wild hominid, although certainly not our species.

A second curious attribute sometimes reported is a type of vocalization attributed to sasquatches in the literature for which there is no known counterpart among the apes, or among humans for that matter. In his 1960 Unuk River, Alaska report of strange vocalizations, Mr. Bruce Johnstone Sr. made reference to what sounded to him like "no language [he had] ever heard." In the 1924 Ostman account from Toba Inlet, Ostman stated: "Then I heard chatter—some kind of talk I did not understand.... They were now standing around me and continuously chattering..." And in reference to the young male taking some artifact over to the adult male, "...he trotted over to the adult male and showed him. They had a long chatter." Although apes occasionally indulge in this form of behavior, there is certainly the implication here that, as in the case with dolphins and chimpanzees, we may not be the only species capable of conversation.

There are more references in Ostman's account to such strange communicative behavior. It is important to note that they do not confirm that this alleged behavior was a form of language. They may be viewed, however, as possibly indicative of such. In most recent reports where a sasquatch was reported screaming, yelling or whistling, there was no other individual sasquatch noted close by. The observed sasquatch was apparently aggressive or fearful, and the vocalization repertoire attributed to sasquatches already includes perfectly suitable calls for such contexts. It seems doubtful, in any such anxious situation, that a sasquatch would care to sit down for a pleasant chat. It is therefore not surprising that we do not have more numerous reports or recordings of low-volume chattering or more complex infragroup vocalizations. There are also several vocalizations that resemble human sounds, particularly whistling and the sound of a woman, child or baby crying and yelling. About the only conclusion we may reach at this point is that alleged sasquatch vocalizations do not fit neatly into either hominid or pongid vocal repertoires.

In some ways, sasquatches appear to be a larger species of solitary, relic hominid, although not of our species. Of course, this is pure conjecture. But it may be fair to say that until the evidence of bones or chromosomes eventually makes its way into the highly respected, if not overly cautious, annals of anthropology, it might be wise to leave open the taxonomic door on the possibility that we may have some new, if somewhat unexpected and unusual, relatives "coming to stay."

Sasquatches as a New Species

One likely hypothesis is that the sasquatch, not exactly ape or man, is conceivably a form so like both that it is not easily grouped with either, belonging instead to a distinct family of its own. This thinking is dependent on the peculiar combination of many apelike and several more manlike characteristics included in reports, and the provision that there are not multiple species of sasquatches with widely differing characteristics.

Fingerprint cataloging for primates is very new, while human patterns are well documented. The dermal ridges of human feet typically show a distinct crosswise pattern of whorls and ridges, while apes demonstrate an axis of ridges along the base of the toes from front to back. Sasquatch dermatoglyphics appear surprisingly different from both ape and man, suggesting a more independent evolutionary path. (This has been suggested in recent forensic work by fingerprint specialist James Chilcutt[16] of Conroe, Texas, and primatologist Dr. Jeff Meldrum in the field of primate fingerprinting has work in progress.) What all this indicates is that sasquatches are, strictly speaking, neither apes nor human, but rather their own separate subfamily, not simply some recent split between living forms or even hybridization of any sort.

Rainforest species in North America are certainly not in a very good position to preserve themselves in fossil form, owing primarily to the acidic nature of the soils. There is only one early fossil hominid known from North America, a small Miocene ape. It is anyone's guess as to how many different forms of prosimians, monkeys, apes and man may have existed, but failed to leave a fossil. Except for

Gigantopithecus, the known fossil candidates just aren't big enough to easily fit the sasquatch. Sasquatches are often reported as more than eight feet in height. But, for the sasquatch, becoming big is logical. Large size would greatly reduce heat loss for ape or man, as would a thickset build. These attributes would also be necessary for defense from predators, since the benefits of group behavior would be absent. A nocturnal activity pattern would allow the sasquatch to move and feed unmolested in any nontropical environment. (Tigers, potential nocturnal predators or competitors, occur in the cold Siberian forests but are relative newcomers there, being evolved in warmer environments.) This behavior would also allow a sasquatch, or its ancestral Asian form, to exist in the proximity of early man, or modern man for that matter, with little interaction between the species. This would be increasingly important if many of the food resources for both species were similar. It is interesting to note in Native ethnographies just how often Pacific Northwest Natives were said to have left a particular food-gathering locale or village for expressly the purpose of avoiding confrontation with sasquatchlike creatures.

In summation, sasquatches may be identified as beings possessing a spectrum of characteristics referred to variously as "apelike," "kind of apelike," "almost manlike" and on occasion "human." Large variations in size, build, color, head shape, tracks, and hair length are all fairly commonplace in human populations—and we are after all a primate too. This is only meant to illustrate that it is possible for a primate species to exist in a continuum of forms.

The evidence, anecdotal though it may be so far, strongly suggests that sasquatches may represent an unknown species intermediate to man and ape, but with many retained apelike traits. There appear to be a few separately evolved traits such as bipedalism, swimming as well as some hypothetically evolved abilities such as simple nest weaving and, possibly, the use of unmodified wood or rocks as "tools." Vocal analysis would appear to resemble the spectrum of the great apes, with the notable exception of complex chattering, which may be more hominid than pongid. In short, what might at first glance to be simply an aberrant ape or relic hair-covered hominid, is likely to be something much more unique and deserving of the attention that science can give. Just how unique will likely be made available for us to determine sometime in the near future, as the database of reported sasquatch behavior continues to expand and physical evidence accumulates.

1. Krantz, 1999, *Bigfoot Sasquatch Evidence*, Hancock House Publishers, Surrey, B.C., pp. 182–93.

2. Lawick-Goodall, J. van and H. van Lawick-Goodall, 1973 "Cultural elements in a chimpanzee community," Symp IVth Int. Cong. Primat. Vol. I *Precultural Primate Behavior* (Menzel, E. W. ed.) pp. 1444–84.

3. Suzuki, A., 1966, "On the insect eating habits among wild chimpanzees living in the savannah woodland of western Tanzania," *Primates* 7: pp. 481–7.

4. Shackley, Myra, 1983, *Still Living? Yeti, Sasquatch and the Neanderthal Enigma*, Thames and Hudson, New York, NY, p. 45.

5. Dixson, A. F., 1981, *A Natural History of the Gorilla*, Columbia University Press, NY, p. 129.

6. Goodall, Jane, 1971, *The Chimpanzees of Gombe*, Belknap Press, Harvard University Press, Cambridge, MA.

7. Dixson, 1981, p. 77.

8. See B.C. report summary, 1965: Nicomen Island near Mission, B.C., [S65A].

9. See Sightings, 1982: reported by Tom B., illustration figure 18, [S82A].

10. Dixson, 1981, p. 76.

11. Green, 1978, *Sasquatch: The Apes Among Us*, Hancock House, Saanichton, B.C., p.150.

12. Kay, R.F., 1987, in *Primate Evolution and Human Origins,* edited by Russell Ciochon and John G. Fleagle, Aldine De Gruyter, New York, Chapter 31, "The Nut-Crackers—-A New Theory of the Adaptations of the Ramapithecinae", p230.

13. Simons, E.L., 1970, "Gigantopithecus," *Scientific American,* 222(1), pp77-86.

14. Krantz, Grover S., 1992, *Big Footprints: A Scientific Enquiry into the existence of the Sasquatch,* Johnson Printing Co., Boulder, Colorado.

15. Coleman, Loren, 1999, *The Field Guide to Bigfoot, Yeti and other mystery primates worldwide*, Avon Books, New York, NY, p16.

16. Meldrum, Jeff, personal communication.

17 Collecting a Sasquatch & Final Thoughts

The rainforests of the world account for many "new" lesser species of mammals each decade. Every now and then, a larger animal is "discovered." The discoveries of the 1900s are surprising: the mountain gorilla, the okapi, the dwarf elephant and the eastern cougar, to name just a few. Over half of all new mammal discoveries are nocturnal primates or prosimians living in rainforests, known perhaps to locals who penetrate those forests, but few others. Even after their discovery, many are so elusive as to appear extinct. The late Oxford zoologist Ivan Sanderson, one of the world's most successful collectors of rare animals, was convinced of the elusive nature of the sasquatch and just how difficult it would be to see one, let alone capture it.

"Why hasn't a sasquatch been caught?" is a question most researchers are still inevitably asked. The question implies that capturing a sasquatch should be easy for anyone with the right equipment. It also implies that someone has funding for such an endeavor. Although several brief attempts were made in bygone decades, there is no one simple method that has proven successful. By the large body of sasquatch reports, this intelligent, nocturnal and reclusive primate is not going to allow itself to be "shot in the dark" nor will it go near a trap of any sort. Following tracks with dogs has proven futile, as the dogs that do not immediately put their tails between their legs are unable to keep up with the maker of the tracks, or else are found wounded or dead. Using attractants has been suggested in the literature. Apart from the possible success in some cases of using predator calls or apples and oranges as bait, no one method has been reported to produce any consistent results. While humans in the back country might certainly consider carrying firearms for safety, it is likely that sasquatches recognize unconcealed arms as a threat, and may be even less likely to allow themselves to move from cover. Aerial infrared searches may locate a careless sasquatch out of the tree cover for a few minutes, but even an equipped helicopter on station will lose the infrared heat signature once the animal has bedded down under the protective canopy of the forest for the day. The forest canopy is so effective in this way that even in aerial search and rescue operations for humans in such terrain, infrared becomes of questionable daytime value. The American or Canadian military may have the resources to mount such an exercise, and may have even been successful as rumored in Alaska, but you and I would never be informed. Civilian scientists would be just as low on the "need to know" security clearance ratings as the rest of us.

Short of a complete body, it is the bone and dental material that science wants most. This is not to say that it is too early to bother with any sighting or track reports or that with the passage of time their value to our knowledge of the sasquatch will not increase. It simply points out that it is now, as it always has been and will continue to be, impossible to expect all scientists to reach a conclusion favoring the existence of a new species at the same time. Instead, the first few to state that there is something new will probably be treated as heretics, then as dogged investigators, next as brave and innovative theorists, and finally as brilliant and farsighted researchers who persistently applied solid investigative logic to the vacuum of ignorance on the subject. It is always, in retrospect, rather comical.

Those who have most to lose by acknowledging the mounting evidence, the professors whose texts suggest "no evidence exists to support this," will quietly slip over to a position of species acceptance in the debate, rather hoping that no one notices them changing camps. Indeed, in the past many new species were discovered only after years of their rumored existence by the most skeptical of the scientific establishment who, when finally cornered by a single type specimen of the species, stated simply, "earlier evidence was inconclusive" (somehow faulting the first and more perceptive researchers).

In defense of the Ivory Tower of academia, it must be stated that the American position has been pressured a good deal by the ridiculous effect of the tabloid media on general public perception of the sasquatch. Nowhere in the world does the popular media approach this subject so hysterically and with such perversion of fact. It is not surprising that so many North Americans are polarized on the subject. The first thing many people with firearms may want to know will be, "Should I kill it or not?" Given the traditional variety of response to this question, it may be assumed that any number of individuals will be quick to resolve this moral quandary with the notion, "if in doubt, shoot first, ask questions later." But most people enjoying the outdoors do not carry firearms. If you enjoy watching wild animals as much as the average outdoorsperson, you might be even less likely to squeeze a trigger. Notwithstanding, just as sasquatches have been reported being shot in the past, such an unusual animal would seem fated to attract someone's bullet eventually, regardless of bioethics or protective legislation. Admittedly, science will insist upon it, as scientific method requires a type specimen in order to pronounce the creature "real" and give it a scientific name. Protective legislation or conservation cannot proceed without the specimen first, as harsh as that may seem to some. It may be some consolation to those placing importance on conservation that producing the first sasquatch will not make it any easier for hunters to get a hold of a second one, given their reclusive nature. In fact, it may be precisely man's spears and bows to which the creature owes its shy and nocturnal habits.

Each year, new species are brought in around the world for taxonomic classification and become discovered. The attention they then receive is not exactly what one could call being left alone. As habitats diminish and more people enter the forests well armed, the time comes sooner or later for all reclusive species to "lay down and be counted." Inevitably, what we may discover with the formal announcement of a new member of the family Hominidae and a rush for museum specimens may be more revealing of the nature of our own species than that of the discovery itself.

The preponderance of documented sightings indicates that, if anything, the behavior of sasquatches is even more retiring and reclusive than that of the great apes; whether this is due to heightened intelligence or simply instinctive behavior is unknown. There remains the possibility that some greater number of reported sasquatch behaviors are learned, as with the apes, either individually or within a family unit. This would certainly increase the variety of behavior patterns as reported observed in different geographic areas. For example, consistently greater aggressive behavior toward man in particular geographic areas may have some basis in what particular sasquatch groups seek to "teach" their offspring. Alternatively, it may be simply an artifact of data collection and circumstance.

Attributed vocal ability greater than apes is also reported with somewhat greater frequency for sasquatches, and there is some audio evidence in support of this, but this too remains hypothetical. In summarizing, most all reported sasquatch behavior is similar to that found in different great apes, actually encompassing behaviors noted in most of the species. Until more definitive behavioral data comes forth, sasquatches should probably be considered as a species similar to the great apes for all practical behavioral purposes, allowing some latitude for those byproducts of vertical posture, a more sophisticated vocal tract and wide-ranging locomotion. Like male orangs, sasquatches are rarely seen, more often heard and just possibly not as solitary as commonly and hitherto supposed. It may be that a single sasquatch seen "in hand" may be worth "two in the bush."

Reclusive, forest-dwelling mammals are difficult to study at best. Add nocturnal or crepuscular behavior and some degree of intelligence to the list of alleged attributes, and it becomes almost impossible. Gorilla researchers at least have large groups of slower-moving individuals to track through the jungle, inhabiting much smaller ranges. The sasquatch, apparently more solitary, is not so plodding or gregarious, although according to some reports, occasionally just as odoriferous. While that and infragroup sounds may help gorilla researchers "catch up" to their subject, sasquatches appear to move faster than humans and catching up is not a field study technique that has proven productive. In fact, there are no such cases.

It now seems all the more likely that reliable data on wild behavior in sasquatches will by necessity come not from lettered professionals examining sasquatch remains, but more likely from coincidental observers. Since few occupations or recreational pursuits take people into the forests at night, the database for sasquatches, at least in coastal Alaska and B.C., will depend on the collection of anecdotal reports such as those in the preceding chapters. All this presupposes that witnesses will continue to be forthcoming in their reported observations. Increasingly, people are becoming dissatisfied with glib newspaper dismissals of reports and have a growing interest in seeing that sasquatch reports are made public. The internet has done much to foster this. It is not likely that the readers of the tabloid media will rush *en masse* to take evening classes in anthropology, but the growing public awareness that "there is something there" will encourage further sharing of new information and material evidence, as well as the continuing availability of such evidence for public scrutiny.

There will most certainly be a "shoot them/save them" scenario. This will not happen because people have long-standing, deep-seated scientific convictions about sasquatches, rather it will happen because most people would simply not have to try to figure out the sasquatch for themselves, and will accept whichever sasquatch label, ape or human, the media chooses on the day the sasquatch is "discovered." In common parlance, it will be, "It's a damn bush ape, shoot it!" vs. "No, it's a wild human, save it!" Finding a sasquatch could happen anytime. Finding out exactly what sasquatches are will take much longer. Anthropology will probably survive it, but none of us will likely hear the end of it.

Today, research on sasquatches is being carried on by both academics and nonacademics. As previously noted, the late anthropologist Dr. Grover Krantz of Washington State carried out a large amount of research on the subject of sasquatches in general, and especially on tracks and reconstruction of foot anatomy, derived from careful analysis of hundreds of tracks. His conclusions that the ankle is positioned much more anteriorly (forward) on the foot than in man seems well supported by the increased flexibility of the sasquatch foot as evidenced in individual tracks within a series. He was also the first anthropologist to point out the definitive ridges and whorls of the toe pads registering in sasquatch tracks. Although retired from university teaching, he continued, until his death in 2002, to be in the forefront of work proposing the acceptance of sasquatches as a species already known, e.g., *Gigantopithecus*.

Also an authority on primate and human anatomy, Dr. Jeff Meldrum of the University of Idaho continues to study casts of tracks matching the sasquatch type and has carried out preliminary field work in the investigation of the hominid responsible for them.

His work with forensic scientist James Chilcutt of Conroe, Texas, on dermal toe prints attributed to sasquatches promises exciting results.

People attempting to photograph sasquatches face the difficult challenge of dealing with a largely nocturnal animal that is intelligent and apparently shy of handheld metal objects such as firearms and possibly anything handheld outside of obvious foodstuffs. Camera flash systems are really inadequate for any nighttime forest photography, unless of course they are infrared, and current built-in infrared light sources for 8-mm camcorders range only to about twenty feet. After-factory accessories project up to fifty feet and have recently become available. In Washington, a growing number of field researchers already familiar with the technologies of law enforcement and surveillance have been able to initiate trials with remote infrared cameras triggered by disruption of a beam or other motion-sensing devices. In the past decade these techniques have proven useful in taking census of Asian tigers, leopards and bears. It is possible that, if they become widely employed, they may have some use in providing photographic details of this elusive creature.

Already, commercially available sensing cameras have been employed by the Washington State Sasquatch Research Group, initiated by Andrew Peterson and George Kelley. This organization utilizes a website (www.wssrg.net) and regional coordinators to allow witnesses and researchers to quickly investigate the site of a reported sasquatch sighting, track or vocalization. They have already refined a system of establishing remote monitors in such areas following reports and are able to work with local witnesses to increase the likelihood of capturing hard evidence of the existence of sasquatches.

The following is an overview of some of the best possible techniques for capturing evidence of sasquatches short of hitting one down with your vehicle or shooting one. (For more on that, refer to suggestions offered by Dr. Grover Krantz in his most interesting and valuable book, *Bigfoot/Sasquatch Evidence*.) They are things almost anyone can do.

1. Should you find yourself looking at a sasquatch from a vehicle, access your camera, which might be kept attached with velcro to the dash or console. After the sighting, stop and note your odometer. Take a quick scan, with a flashlight if necessary, to

look for tracks, which can be marked with cardboard, an old floor mat or some marker that will not blow or wash away. A small piece of surveyor's tape fastened to a branch or placed beneath the edge of a rock will aid in reexamining the site for tracks, hair, etc., at a later time.

2. If you are out in the bush on foot with one or more companions and you believe you are observing a sasquatch, one thing you should not do is take your eyes off it to discuss it with your friends, who are probably asking "Did you see that?" This is important because most sightings, as with any wildlife sighting, last only seconds. Whether you have photographed the creature or not, you may also choose to follow the steps outlined in the first step regarding marking your position (especially if you have a GPS), checking for tracks, etc. It is probably a good idea to photograph the spot for future reference if you are carrying a camera. Hair found on branches, tree trunks or barbed wire should be collected and placed in a bottle containing an alcohol solution for hair with root follicles (possible DNA sample) or saline solution for hair shafts without roots. Forward the sample to any previously named researcher with scientific credentials.

3. Make brief notes of your encounter, the location, the type of vegetation, weather at the time or anything else out of the ordinary you may have noticed at the time of your sighting. Mark the location on a map should you forget the exact spot.

4. Unless there seems to be an ongoing threat to life or property, do not bother calling the local police, fish and wildlife, or anyone else who is probably not really interested.

5. Call or e-mail a discreet research group, and submit your information to a database. You should expect confidentiality, and confirm the organization's standard policy on same. In this way, interested scientists or researchers will be able to share data and further knowledge on the species. Your factual information regarding such creatures will make a contribution to science.

One British Columbia forestry professional, quietly claiming to have some experience with sasquatch reports in northeast B.C., is forestry biologist Regan Dickinson. He has sorted through numerous sasquatch reports from individuals in the Dawson Creek and Ft. St. John areas, and advises the following common sense for individuals seeking sasquatches.

"Packing a gun is a good idea, but, again, the sasquatch will stay hidden. Packing bear assault spray is recommended, since this stuff will drop a grizzly, and if the sasquatch attacks, spray it into his large eyes. An air horn may get some response out of it if you want to hear or record calls. Respect for the sasquatch is important and one should avert their gaze if possible. And you may try to move toward it in a nonthreatening way, and just look at it from a good position. Sit down and just watch it until it gets bored and moves away, but it may outlast you in this game. Try to take some camera shots. If it threatens loudly, move out of the area as promptly as necessary."[1]

While Mr. Dickinson's recommendations may seem casual to some, they are in fact well thought-out suggestions based on field safety guidelines in an area where grizzly (and sasquatch) reports are not infrequent. Of course, for individuals eschewing digital or celluloid shots in favor of a heavier format (such as lead), there will be little need for the above behavioral observations and a much greater need for a knowledge of anatomical dissection skills.

Specific information on other useful field techniques and equipment can be found on the website of the continent-wide Bigfoot Researchers Organization (www.bfro.net), which lists reports in each state and province. BFRO also details video- and audio-recording equipment, as well as casting materials and techniques. Likewise, the Washington State Sasquatch Research Group offers discreet help and shares reports. In Oregon, the Western Bigfoot Society (email <raycrowe@aol.com>), and in California, Bobbie Short's website (www.sierra2.net) also file reports and coordinate research. British Columbians might contact Dr. John Bindernagel (email <johnb@island.net.ca>). Some field investigators may discover their experiences match those of others and together form one piece of a larger puzzle. It is a cryptozoological puzzle comprising thousands of items of reported evidence. With each piece of understanding we are that much closer to seeing the emerging picture of the sasquatch. With burgeoning academic interest and accelerated field research occurring on a widespread basis, the missing pieces are becoming fewer. Since the phenomenon of the sasquatch shows no indication of fading away, it is just possible that the next individual to step forward with some information regarding sasquatches may be the person who holds the last piece to the puzzle.

There are four main conclusions that may be drawn from reviewing everything that has been said. They are not contradictory and may be stated as follows.

1. It is certainly possible for a primate as large as the sasquatch to exist. There is at least one extremely large hominid, estimated variously at eight or more feet in height, known as *Gigantopithecus*, which, according to physical anthropologists, survived from seven and a half to until at least half a million years ago. Given a bell-shaped curve for the existence of the living animal well before and after the earliest and latest fossils, the extrapolated existence of such a creature would easily fall into the present time. No anthropologist can prove it was not erect. The absence in North America of other fossil forms such as the African *Australopithecus robustus* or the Asian, European and African *Homo erectus* is not proof that, in small numbers, those other species did not continue to migrate elsewhere and/or evolve into larger forms without leaving fossil evidence.

2. The Native folklore of western coastal North America has numerous names and descriptions of just such a large, nocturnal, hair-covered humanlike being reputed to inhabit the forests and mountains of the Pacific Northwest. It is noted that, although many aspects of traditional Native belief systems have, through western disbelief and ridicule, been replaced by western culture, many of these Native beliefs in such a creature persist to this time.

3. There exists a body of documented reports from individuals on the B.C. and Alaskan coasts, among other areas, of the existence of large, hair-covered hominids. They appear generally reclusive and are observed by man just as often at night as in the daytime, which suggests somewhat nocturnal behavior for that species given the far greater number of people about during the

daylight hours. The reports vary in size and color, as they would for human, but share the more apelike characteristics of heavy build, hair-covered body, flat feet and generally apelike vocalizations.

4. Larger-than-human tracks displaying a typical shape with wide heels and no claws or human arch have been found over a broad area of the Pacific Northwest and elsewhere in North America, often in association with sightings, unexplained vandalism or, on occasion, vocalizations of unusual volume and form. The tracks themselves are often found in very remote places where man's presence is almost nonexistent. The pattern of dermal ridges on the toes, themselves difficult to hoax, has been found to be of a type intermediate between man and known apes. Long hair of a type previously unknown for North American mammals has been collected and typed by Dr. Henner Fahrenbach.

None of the conclusions in each chapter are by themselves greatly significant. When viewed together, however, they suggest something is certainly there and is, in some ways, already well known. Any hard biological evidence collected by the average person may, with repeated luck, find its way to one of the researchers previously mentioned who will have the good sense to know what to do with it. That is not really the problem. The main impediment to further knowledge regarding sasquatches at this time, is the relative lack of familiarity the average observer has with the appearance, tracks, sign and behavior of the creature, and in knowing where to forward a report. Even after most sightings, the question still remains, "What on earth was that?" Hopefully, the appendix "Field Guide to the Pacific Coastal Sasquatch," will be of benefit in familiarizing anyone new to the subject of sasquatches with what they might expect to see when spending any time looking at tracks, signs or just possibly a sasquatch itself...

1. Dickinson, Regan, Western Bigfoot Society newsletter, *The Track Record*, ed. Ray Crowe, Hillsboro, Oregon, October, 1998, no. 82, p. 18.

Appendix One

Field Guide to the Pacific Coastal Sasquatch

Physical Description

Attributes
Sasquatches are large, forest-dwelling hominoids. They resemble an upright ape in the torso with no visible neck, are taller than humans, have broader shoulders and pelvis, and have longer legs adapted to a bipedal gait. They are generally reclusive, nocturnal and fairly solitary.

Height
Adult males: 7 to 9½ feet, occasionally larger, averaging 7½ to almost 8 feet; females: 6 to 7½ feet; adolescents: 6 to 7 feet.

Weight
Males: 500 to 1,100 pounds; females: 400 to 700 pounds.

Tracks
Males: 14 to 20 inches (average 14 to 16 inches), width 6 inches or more; females: 12 to 16 inches, width 5 inches or more; adolescents: 7 to 8 inches or more; infants: 5 inches.

Body
Massive in adult, with deep thorax and wide shoulders (disproportionately wider than a human), head elongated, some individuals with a conical cranium or sagittal crest giving a pointed appearance, females less so. Both sexes show a flat face, distinct brow ridge, eyes appear sunken; broad zygomatic arch and mandible. Lips not

everted. The nose is variously broad with nostrils projecting anteriorly, may be more pronounced and convex in some individuals. Ears usually concealed by hair. Arms massive with tapering biceps and forearm slightly longer proportionately than in human. Neck extremely muscular with head held somewhat forward, appearance is one of no neck or "bull neck." Adult females may show visible breasts and a broad pelvis. Lower extremities are long with proportionately robust musculature. Knees do not fully extend in normal ambulation. Ankle anatomy is more anterior on the foot than in humans, giving the impression of a longer, more projecting heel. Hands and feet are long and broad, somewhat disproportionately so, the thumb being slightly smaller in relation to the hand than in man. Palms are relatively long; the thumb is only moderately opposable. The feet are without an arch in normal weight-bearing, but all extremities are extremely flexible, especially in extension at the wrists and ankles. Digits are also extremely flexible; both fingers and toes may abduct to a great degree. The sole of the foot, in adults, may be several inches thick.

Pelage
The entire body is hair-covered, with the exception of facial, palmar and plantar surfaces. Skin color is grayish brown to black in older individuals. Female breasts are hair-covered with exception of the areolae. Overall hair length is three to five inches, occasionally longer on head, shoulders and upper back to a length of eighteen inches or more. Hair lacks a continuous medullary canal and is worn round or split at distal extremity. Hair color usually uniformly black, dark brown, variously reddish brown or beige, occasionally grayish, rarely white. In older individual males there may be whitish areas on lateral jaw or neck. No bicolor patterns are noted in the Pacific Northwest.

Locomotion
Walking, running, jumping, climbing of both trees and rock faces, swimming both on surface and underwater (arms held in front or at sides, leg motion swimming described as a "frog kick").

Manipulation of objects
Sasquatches manipulate objects such as branches, logs and rocks, as

well as manmade objects, with a palmar grasp. There is no documentation to refute partial opposition of thumb. Throwing objects such as rocks, branches and logs is all with underhand or side-hand motion.

Activity pattern

Nocturnal or crepuscular (dawn-dusk active) with daytime activity minimal. Sleeping on ground in prone fetal or side-lying fetal posture within cover or thicket, rarely with little or no cover, preferring dense forest during daytime, sometimes a nest with or without a canopy, occasional use of rock shelter and possibly cave in some instances.

Diet and Subsistence Patterns

Sasquatches appear omnivorous and well adapted for eating leaves, conifer buds, the inner cambium layers of bark, conifer cones and seeds, tubers, fruit and berries as well as various species of edible seaweeds. Tubers are occasionally washed and left in stacks, and very large caches of piled berries have been noted, presumably for drying or later retrieval. Sasquatches have been observed carrying armfuls of seaweed, species of grass, conifer branches, as well as animal carcasses.

Animal diet is obtained by foraging, scavenging and predation. Sasquatches have been observed clamming, usually by hand, but occasionally with simple tools such as large sticks. Piles of shells attributed to sasquatches have been noted in association with tracks confirming strong swimming or diving abilities. Other marine species documented include mussels, oysters, scallops and crabs, all available in the intertidal zone. In estuaries, rivers and lakes, salmon and char have been associated with feeding sasquatches. Sasquatches have been observed wading and diving for water lily tubers, and seen at close distances chewing the ricelike tubers of chocolate lily or "Indian rice." Skunk cabbage and devil's club stems and roots have been noted in partially consumed states in association with sasquatch sightings and tracks, forcibly pulled from the ground at their stem bases.

Smaller mammals such as voles, mice, hares and rabbits are documented in sasquatch diet, and approachable species such as porcupines, grouse and ptarmigan may be obtained where available.

There exists documentation of killed waterfowl being carried by a sasquatch in association with the use of a long stick, presumably a hunting tool. Larval insects have been identified in association with observed sasquatch foraging behavior through old logs or trees. Large game such as blacktail deer have been documented as prey species, obtained by both ambush and pursuit. The method of capture where witnessed is usually a manual blow to the head, neck or body with bimanual snapping or twisting of the neck. Smaller game is reported often as being flailed against an immobile object such as a large tree. This behavior has been noted frequently where domestic canines have pursued and harried sasquatches, but there is no evidence of the dogs being consumed. In the cases involving ungulates, the carcass is sometimes torn into portions, occasionally areas of skin with fur are removed and left. Occasionally only high-value organs such as liver and heart are found removed, and carcasses have been documented occasionally left hanging in trees. Scavenging of other animals' kills has been recorded, as has witnessed theft of human-killed game. Large ungulate carcasses (appropriated from predators' kill sites) up to the size of elk or moose have been reported in association with sasquatch tracks. Carcasses have been found cached, buried and camouflaged in association with sasquatch reports. Black bears have been observed in combat with sasquatches, the outcome unrecorded. It is not known for certain that black bears, either hibernating or otherwise, are a prey species for sasquatches, but it seems likely. No such documentation exists for grizzly or larger brown bears.

Caches of stored plant materials have been found in possible association with sasquatches, and it is possible that meat may be similarly cached, either in a cave shelter, in cool water or, if carried to a sufficient elevation, in snow or ice.

A noninclusive list of edible northwest coastal plants and animals available for utilization by sasquatches is presented here seasonally as follows: marine life such as shellfish and abalone, limpets, gumboots and other chitons, herring eggs, crab, salmon, trout and char; various seaweeds such as ribbon seaweed, black seaweed, sea lettuce, bull kelp, brown seaweed, goosetongue and beach asparagus; various foliage such as deciduous leaves, conifer needles and seeds, including hemlock, fir and pine; cambium of hemlock and spruce; grasses; various nonpoisonous forbs such as leaves and stems of fireweed, wild celery; various fruit and berries such as salmonberry, huckleberry, cranberry, currants, rose, elderberry, salal and wild crabapple; various corms and tubers such as wild onion, camas, chocolate lily, Indian rice, pondweeds; fern roots; wild potato; skunk cabbage; devil's club (cambium, lower stalks and roots); various rodents such as voles, marmots, lagomorphs and porcupine; various ungulates such as deer, mountain goat, occasionally elk and moose; various migrating or nesting waterfowl; various game birds such as ptarmigan and grouse; scavenged wolf or cougar kills.

Fecal material (scat) is described as similar to a bear's; it is ropelike with a diameter of two to four inches, but often in a much larger mass, occasionally as a huge amorphous pile up to a volume of several liters.

Social Behavior, Aggression, Territoriality

Sasquatches have been documented as being solitary great apes.[1, 2] By reports they appear to be somewhat wide-ranging and retiring. Individuals assumed to be males are seen more often than (assumed) females by a ratio of approximately forty to one. Adult males are generally solitary but have been noted in family groups with females and infants. Infants are almost never seen without an adult female. Adolescent males and females may be indistinguishable and are approximately the size of a large man. Smaller, more gracile individuals from five to six feet in stature are presumably adolescents or older juveniles.

Sasquatches have been reported as usually reclusive but also, on occasion, quietly observing man. Theft of human property, food and artifacts have all been widely noted, in some cases reciprocating with items from the natural environment. Solitary play such as hopping and somersaulting has been reported in juveniles. Wrestling among adult or adolescent male sasquatches has been documented. This may represent play or more ritualized behavior. Swaying the tops of trees has been noted by several observers, as has "playing" with some human toys and tools left out overnight. This list includes toy balls and tin dippers at the diminutive end and extends to axes, furnaces, picnic tables, vehicles and logging rigging at the upper end. Manipulation of vehicles, campers, trailers and occupied human structures is also widely documented and may represent anything from play and curiosity behavior to serious threat display.

Curiosity or mild threat display has been documented in the tossing of small stones or pebbles on tents, campers, etc. Increased aggression is noted in the form of whooping, screaming and the throwing of logs or rocks. There is documentation of a rising scale of anxiety and/or aggression in apparent ritual threat display which may culminate in slapping trees, chest beating and the breaking of vegetation up to tree size. In extreme cases sasquatches are reported to react to invasion of critical personal space (or to being wounded) with violence toward the intruder. While such actions have not been commonly reported or their veracity ascertained, it should be deemed possible and caution exercised.

Territoriality in sasquatches is assumed at least in some form, and the marking and defense of such territories is hypothetically indicated by twisted saplings, broken trees and possibly rock cairns. Adjacent or overlapping female territories are assumed to be up to fifty or a hundred square miles, with male territories up to four or five hundred square miles, possibly less in favorable habitat. Females and adolescent males presumably out-migrate as environmental factors dictate. Rock cairns or bent trees marking the passage of males may be left for females and adolescents to follow. Dominant males presumably offer protection from large predators or competitive males. Female territories are presumably more distant from areas of human habitation and suited for occupation year-round. Sasquatches usually seen are more often young adult or subadult males.

Mating, Gestation, Birthing and Development

Little is known of breeding behavior although nocturnal primates generally favor a harem system of mating. In Alaska, a near full-term pregnant female has been reported observed in October, which given the nine-month gestation period common to man and the great apes, would hypothetically place mating in March and birthing in November. Although this hypothesis would accord well with fewer documented sightings in the months of February, March and April, the low frequency of early spring sightings may be an artifact of low human activity and observation at that time of year. As in other hominids, there may be no particular accordance of mating with any particular season or of birthing with any particular season. Nest structures, especially those documented as having a canopy, and close maternal protection may provide the only necessary shelter required of infants in the first year of life. Some field reports suggest that infants may be born with little hair. Reports of infants with mothers occasionally mention the carrying of offspring under one arm, and other reports of solitary infants or maternal-infant pairs indicate infants developing walking skills by the time they have reached three feet in stature. Based on observations of females with young, it has been surmised that one female may care for one infant for seven or eight years before again conceiving, but this is conjectural.

Observation of paired adult-infant and adult-juvenile groups of tracks suggest a minimum of three to five years between births.[3] Maturation of females has been estimated at approximately ten years of age and at a stature of approximately six feet, when females attain development of secondary sex characteristics.

Adulthood may extend into the third decade and, based on research on the great apes, forty-five years would be a maximum. Average lifespan is probably thirty to forty years. Occasional individuals may perhaps attain a life span of fifty or, exceptionally, sixty years, but this is speculative.

Daily and Seasonal Activity Patterns

Sasquatch reports strongly suggest nocturnal or at least crepuscular activity patterns, but reports are skewed by diurnal human activity. The diurnal optical acuity of sasquatches has been suggested as being similar to that of humans. Documented nocturnal eye-shine, whitish red or red does not suggest the presence of a tapetum (reflective pigment layer on the retina), but this is not known.

The prevalence of red as the reflected eye-shine color does suggest either a very richly vascular retina or possibly increased numbers of rod cells to facilitate nocturnal vision. Nocturnal vision may be in the order of two to three times greater than man. Numerous nocturnal reports involving headlights and flashlights, illuminated onto the faces of sasquatches, include mention of aversive movement of the head and arms, as in man. Vision in extremely low light levels is not known, but not likely to be very great. This may correspond with stationary foraging or resting during deep darkness with periods of increased travel and predation of game during dawn-to-dusk hours. Sasquatches are least observed at midday and, when observed resting or rising from a resting position, such observation is particularly during daylight hours. Using estimates of the comparative activity and presence of human observers at day and at night, the distribution of night to day reports suggests an animal almost wholly nocturnal in its behavior, with late evening to midnight and early morning until just after sunrise representing the times when human-sasquatch interactions are most likely to occur. Interactions with stationary humans in sasquatch habitat are also likely in the interval from midnight to dawn.

Observation of sasquatches at higher elevation in winter, in both the subalpine and alpine ecosystems, is noted more rarely and it is generally assumed that this behavior is related to migration from one part of a territory to another. Since this is poorly understood, it may also be that predation on particular winter-available species, increased need for plant foraging or scavenging for winter-killed carcasses are factors. There is no documented evidence that sasquatches migrate long distances from any home area in winter.

Vocalization, Communication and Intelligence

Vocalizations commonly attributed at a distance to sasquatches include screams, whistles, yells, roars and crying noises. Volumes of up to 120 decibels are reported. These are all likely locator or threat displays. Cries may be wailing, as in a siren, or crying not unlike that of a baby, child or woman. A whooping or ululating call is also reported. At closer range, sasquatches have been heard to grunt, chatter, squeal, moan, mutter, bark, laugh, cough, whinny and yelp. The similar vocalizations of cougar, porcupine, ptarmigan and all of the local species of owls may not be easily distinguished from sasquatch vocalizations except by lesser volume. The most easily identified sasquatch calls would include whooping, shrill whistling, childlike crying or loud humanlike yells, all in the absence of nearby humans.

The significance of unintelligible chattering, muttering and longer, more complex chains of vocalization does not preclude the possibility of a simple form of "speech" as a means of short-range communication. Loud tree-rapping or wooden striking sounds are attributed commonly to sasquatches in historical and modern reports. This sound may be intended for the purpose of a warning to other individuals or locator behavior. Answering rapping is documented.

Apart from vocal ability, the intelligence of sasquatches may be viewed as roughly equivalent in some ways to that of a young child. Aptitude for problem solving is apparently high, similar to that of an orang mechanically and to that of chimpanzees or bonobos in terms of simple tool using. Sticks, clubs, branches and rocks are commonly reported in association with sasquatches and the function of tool using in a problem solving situation would likely be greater than most of the higher great apes. There is some indication that at least certain individuals are aware of leaving tracks and generally avoid doing so, unless for the possible purpose of marking their presence to man or other individual sasquatches. Mimicry of human behavior is occasionally noted.

Habitat, Range and Population

Along the coasts of Alaska and B.C., sasquatches typically inhabit areas of temperate coniferous forest at elevations from sea level to the subalpine. Reported coastal range is from Northern California to Alaska. There are more isolated reports from as far north as the Kenai Peninsula, the Kuskokwim and Yukon Drainages, as well as interior Alaska. In southeast Alaska, locations near the ocean, areas of dense old-growth or mature second-growth forest mixed with muskeg/shore pine associations appear to be favored core habitat. Access to beaches, tidal flats or rocky shoreline with undisturbed local populations of shellfish appears significant. Habitat combining the above with a high distribution of muskeg-shore pine ecosystems is also favored, especially during winter months.

Inland, sasquatches apparently inhabit remote low-lying river valleys with mature timber and drainages supporting high populations of anadromous salmonids. Areas of dense salmonberry at low levels, and huckleberry-blueberry at both lower and higher elevations also appear to be preferred habitat. The occurrence of sasquatches in the alpine and glacial ecosystems would appear to be seasonal or limited to the passage of individuals from one drainage system to another. Tracks are often found along river systems at low elevation and ridge lines at higher elevations, indicating corridors of travel.

The range of individual animals is speculative and may be influenced by the frequency of movement within a range. This range may be upwards of 400 square miles and overlap with ranges of adjacent individuals. The usage of a number of individual, widely separated ranges or food gathering locales has been theorized, with the passage of animals along a familiar corridor in between. The coastal ecosystems are supportive of a high biomass and 100 square miles of this type of habitat could be viewed as easily carrying one sasquatch. Interior ranges and subalpine/alpine areas would require two to three times that area. Males ostensibly range further. Range for a dominant male might include 400 to 500 square miles embracing a variety of all above ecosystems. Female territories are apparently more remote from human centers of population and females with young are not commonly observed. The range of a female would be in the order of 100 square miles, optimally embracing all of the above ecosystems. Less dominant mature males and subadult males appear to transit greater distances. It is considered that periodically, at intervals of eight to fifteen years, individuals may change their range completely, possibly relocating in an unoccupied range.

Sasquatches appear to be widely and sparsely populated. A tentative and approximate ratio for population density has been put forward by Bindernagel, based on ecological considerations and British Columbia sasquatch track and report frequencies. This model utilizes known coastal population clines for the black bear, *Ursus americanus*. The relative population ratio hypothesized for the two species within the same habitat is approximately 200 black bears for every sasquatch.

Given this ratio, an approximate population estimate for sasquatches on Revillagigedo Island, Alaska, would be in the order of twenty to thirty sasquatches. A similar estimate for Prince of Wales Island might be 120 sasquatches.

Using the coastal mountains as a boundary, it might be expected that southeast Alaska, Vancouver Island and the southeast B.C. coastal mainland may support approximately 2,000 sasquatches. Bindernagel's ratio model may be conservative, the population may be many times that or possibly less, but the latter is doubtful for genetic reasons. Noting that these coastal areas yielding reports of sasquatches are all continuous or at least adjacent and the apparent ease with which sasquatches locomote in all terrain at brisk speeds, the maintenance of the sasquatch gene pool does not appear to be a very great factor. While reported variation in individual sasquatch size and color might indicate some degree of genetic drift occurring, there is certainly no reason to conclude that the sasquatch is not well adapted to existing with what we might term a sparse population density. There is no reason at all to conclude it is in any way endangered. Even in many areas of widespread logging or multiple land-use management there have been no great reductions in the frequency of sightings and other reports.

This brief guide to the Pacific Northwest sasquatch of southeast Alaska and coastal B.C. is intended to provide an approximate outline of what may be typically observed. Individuals using this guide should bear in mind that their own observations and intuition may produce far greater success in detailing sasquatch anatomy and behavior. Note-taking and collection of data in the field is recommended, since it gives researchers a framework and common language for sharing collective sasquatch data. As well, sharing of research data will allow individual investigators to obtain information and local assistance from other more experienced individuals and cooperative groups of researchers. There will be no *"Sasquatches in the Mist."* Anyone now claiming to have "lived among the sasquatches" without photodocumentation must be viewed as circumspect, if not "circum-the-bend." With the presentation of the first anatomical specimen of a sasquatch, anatomy will quickly overtake behavioral and ecological knowledge of the sasquatch, emphasizing how little we really understand of this primate's habits and requirements.

What can be searched for successfully by most anyone is knowledge regarding the behavior of sasquatches. Newer photographic and surveillance technologies may offer promising results to experienced investigators and organized groups of researchers, but not necessarily for individuals. Rather than loading rifles or engaging in persistence hunting to run down tracks, it may be just as profitable for individual researchers to proceed slowly, learning from the observation, recording signs and behavior represented, then contributing to a database.

Observation is simply a tool for biologists, naturalists and those who enjoy exploring the diversity of nature and discovering its secrets. Observation without basic knowledge of what one should expect to find can be frustrating. By sharing the consensus of knowledge and current research concerning sasquatches, it is hoped that the preceding chapters may be of some assistance to those wishing to study this fascinating primate. With a broadening base of field investigators, it is likely that cryptozoology—the scientific pursuit of unknown life forms—will soon have something to offer quite close to home. We only have to look.

1. Green, 1978, *Sasquatch:The Apes Among Us*, Hancock House Publishers, Surrey, B.C., p. 449.
2. Krantz, Grover, *Sasquatch/Bigfoot Evidence*, Hancock House Publishers, Surrey, B.C. p. 164.
3. Fahrenbach, H. W., in "Sasquatch: Size, Scaling and Statistics," *Cryptozoology*, Vol. 13, pp. 47–75.

Appendix Two

A Summary List of Southeast Alaskan Reports

It appears that sasquatches inhabit virtually all areas of southeast Alaska. Just how populous sasquatches might be on the coast is difficult to determine, but there seems to be no reason to assume they are not far-ranging. To gain some appreciation for the similarity and geographic continuity of sasquatch reports, we can compare the annotated summaries of documented sasquatch sightings in southeast Alaska with those that follow along the British Columbian coast to the south. The following is a summary list of reports of sightings recorded for southeast Alaska. They are, for the most part, from areas closer to Ketchikan and central Prince of Wales Island (abbreviated P.O.W. Is.).

The report codes for southeast Alaska are provided as letter/number prefixes in brackets, e.g., [OS00]. The initial letters are used to signify the type of report and its chapter:

OS: Old Stories from the Coast
B: rock or log throwing, vandalism, etc.,—Sasquatch Behavior
M: bones—Traces of Hair and Missing Bones
N: nests —Nests and Other Signs
S: sightings —Sighting Reports
T: tracks—Big Footprints in the Panhandle
V: (or VO) vocalizations—Reports of Possible Sasquatch
                    Vocalizations

The middle two digits provide a reference for the year of the report, e.g., [S52A] would be a 1952 sighting, while the final letter A, B, C, etc. is used to distinguish reports of the same type occurring in the same year. Alaskan codes are enclosed within [square] brackets while British Columbian reports in the following chapter are enclosed within (round) brackets. Both B.C. and southeast Alaska reports are coded on the maps over their approximate location of occurrence, as well as in the two summary lists. Alaska reports in the main text also include codes for direct map reference.

[OS00] 1900: Thomas Bay, AK, account of hair-covered creatures chasing man, other mysterious incidents from Harry Colp's "The Strangest Story Ever Told." [map2]
[OS09] 1909: Frank Howard's Malaspina Glacier story, "The Beast in the Glacier." [map1]
[VO21] 1921: Claude Point, north Revilla' Is., Jack McKay hears three loud humanlike calls, no one there. [map5]
[OS20] 1920s: Chinatown, Annette Island, rocks thrown at cabin at night by "just some old monkey," large tracks seen the next day, Mrs. N. F. report. [map5]
[OS28] 1928: Hecata Island, a Mr. Watson driven off prospecting claim by numerous three-and-a-half- foot-tall apelike creatures. [map3]
[VO31] 1931 (approx.): Claude Point, north Revilla' Island, Jim Pitcher reported hearing three humanlike calls, but "no one there." [map2]
[NS30] 1930s, 1940s: Unuk and Chickamin Rivers, Bruce Johnstone Sr. reports limbs four inches thick twisted off in various locales, not apparently done by moose or bears. [map3]
[S42A] 1942, May: Bob Titmus reported to John Green that somewhere near Wrangell Narrows, while going on deck of a small ship late one evening, he saw on the beach, close by, a seven-foot-tall erect creature, heavily built. [map2]
[VO44] 1944: Claude Point, north Revilla Is., Ruth Jackson hears three humanlike calls, no one there. [map5]
[VO44] 1944 (approx.): Claude Point, north Revilla' Is., B. Johnstone camped, hears a single loud long humanlike call, no one there. [map5]
[B40A] mid-1940s: south P.O.W. Is. near Point Chacon, Mrs. V. M. reports log being thrown from bush at her and her family. [map3]
[B40B] late 1940s: Thomas Bay, AK, Mrs. J. M., second-hand report of son-in-law and friend experiencing huge snowballs lobbed at boat. [map2]
[B40C] late 1940s: near Hollis, P.O.W. Is., W. K. reported caterpillar turned over two consecutive nights. [map4]
[MB40] late 1940s: near Annette Is., AK, J. M. reported giant skeleton with thigh-bone three to four feet long. [map5]

[V48A] 1948 (approx.): Bakewell Creek, Smeaton Bay, AK, Mr. Bruce Johnstone Sr. hears two creatures chattering behind tent at night, "not porcupines." [map3]
[V48B] 1948: California Head, Revilla Is., Mrs. B. B.'s second-hand report of hunter hearing noises "not a bear." [map5]
[MB48] 1948: Bartholomew Creek, Smeaton Bay, east Behm Canal, AK, two men report finding "jawbone larger than a man's." [map3]
[VO50] 1950: Soda Bay, southwest P.O.W. Is., Al Jackson and father hear repeated screams from boat, father stated "not a bear." [map3]
[S51A] 1951: Klawock, west P.O.W. Is., Miss M. reported vandalism around house, vocalizations, hair, reddish and black sasquatches seen. [map4]
[S52A] 1952 (approx.): Klawock, western P.O.W. Is., Mr A. K. reports vandalism, vocalizations around same house as [S51A] by sasquatch. [map4]
[VO52] 1952: Wild Man Cove, Dora Bay, southeast P.O.W. Is., Bruce Johnstone hears loud yells circling him by muskeg. [map4]
[S55A] 1955: Rudyerd Bay, East Behm Canal, second-hand report, a hunter alleges watching a ten-foot, black sasquatch stampedes goats down toward them. Nineteen-inch tracks reported. [map3]
[S55C] 1955: North Tongass Rd., Ketchikan, six youths report a ten-foot-tall black sasquatch chase deer across road in front of them. [map5]
[VO56] 1956: Bishop's Island, Unuk R., AK, three men all listen to wailing, chattering, circling camp for one hour. They called back and forth with it. [map3]
[VO58] 1958–60: Klawock, P.O.W. Is., AK, reported loud yells and screams, no one thought were bears. [map4]
[S50A] 1950s: Klawock, P.O.W. Is., AK, Mr. A. K. reports damage to roof, screams, bear traps sprung with sticks overnight. [map4]
[S60A] 1960: Metlakatla, "Erroll" and group of fishermen see tall sasquatch stand in water by dock, then swim underwater. (John Green, *Sasquatch, The Apes Among Us*, p. 432.) [map5]
[S60B] 1960: Ohmer Creek, south of Petersburg, Paul M. sees legs of sasquatch by flashlight just yards away in creek at night. [map2]
[N60A] 1960: Connell Rd., Revilla Is., AK, Bruce Johnstone Sr. reports fresh nest in summer with woven canopy. [map5]
[S66A] 1966: Ward Lake, AK, Ms. S. H. and two other women see sasquatch by tent, relocate to picnic shelter across lake and are later followed, sasquatch steals ax. [map5]
[B66A] 1966: Ward Lake, AK, D. Y. and five others in car picked up by rear bumper and car shaken. [map5]

[S68A] 1968 or 1969: Wrangell Narrows, Kupreanof Is., five boys hear rapping, one sees brown form. [map2]
[S67A] 1967 or 1968: north of Tombstone Bay, Portland Canal, AK, Tom B., with two men, reports watching dark sasquatch digging with stick on beach. [map6]
[S69A] 1969: east of Wrangell, above Bradfield Canal in mountains, T64s, R88e, sec23, ne., Mr. J. W. Huff and man watch sasquatch observing them. [map3]
[V70A] 1970 (approx.): Stikine River, above Wrangell, Harvey Gross with others reported three loud screams, "not cougar." [map3]
[N72A] 1972: Connell Road, Revilla' Is., AK, M. J. Hert reported fresh summer nest, eight feet in diameter, above logging road. [map5]
[B74A] 1974: Ward Lake, AK, L. W. and four others in car picked up by rear bumper and shaken. [map5]
[V74A] 1974: north of Swan Lake, Revilla' Is., A. J. and others heard a loud screaming yell, "not bear or human." [map5]
[S74A] 1974 (approx.): Spacious Bay area, west Behm Canal, AK, tugboat man R. W. sighted sasquatch on beach. [map5]
[T74A] 1974, April: Thomas Abbott reported in 1999 that ten miles north of Haines, he and another man had seen one eighteen-inch manlike track and hanks of equally long reddish hair along with copious blood and wolf tracks. [map1]
[V74B] 1974: Saltery Cove, Skowl Arm, east P.O.W. Is., Mrs. J. M. said husband heard scream, saw one big track. [map4]
[V75A] 1975: Saltery Cove, Skowl Arm, east P.O.W. Is., Mrs. J. M. and family all heard series of loud screams. [map4]
[S75A] 1975 (approx.): Chomondoley Sound, P.O.W. Is., engineer R. S. and daughter M. Z. saw reddish creature with long legs moving, "bent over" on beach, viewed from sailboat. [map4]
[V77A] 1977: Granite Basin, six miles north of Ketchikan, Revilla' Is, Harold Omacht reported series of very loud calls, "chuckling" sound, "not a ptarmigan." [map5]
[S78A] 1978: Saxman boy K. S. reported to friend that he watched furry manlike creature drink from cupped hand at nearby creek. [map5]
[V78A] 1978 or 1979: Last Chance Campground, Revilla Is., Rod Stockli and others, heard long, loud scream, "kind of like an elephant trumpeting, only louder." [map5]
[T79A] 1979: Gem Cove, east side Carroll Inlet, Revilla Island, Mr. A. S. finds three sixteen-inch tracks, no claw marks, one and one quarter inch deep in muskeg, his 315-pound frame registers only one quarter inch deep when jumping on one foot. [map5]
[B79A] 1979: Cedar Pass, south of Sitka, H. A. reported huge hairy hand

and arm reach through porthole, lifting up his arm for a moment. [map2]

[S79A] 1979 (approx.), fall: late "Tink" Durgan reported to Terry Wills and others that at Hidden Inlet, south tip of AK panhandle, he had seen a seven-foot, black, man-shaped creature watching him from shore as he aimed his rifle at it. It walked away on two legs. [map6]

[S70A] late 1970s: Yakutat man, Steve Johnson, reported to Fred Bradshaw, berry pickers shot at seven- foot sasquatch, the creature returned in following years. [map1]

[S70B] late 1970s: mile 13 north of Ketchikan, Mrs. K. W. reported a seven-foot, dark sasquatch standing beside highway. [map5]

[S70C] early 1970s: Annette Island, north of Metlakatla, hunter E. B. reported seeing an upright, hair-covered figure running across logging road "faster than a deer." [map5]

[S80A] 1980: Hessiah Inlet, west P.O.W. Is., Mike Stough, Wrangell, reported two tall black creatures digging on beach, tracks found by men. [map3]

[S80B] early 1980s: Tina D. reported in summertime seeing a small three to four foot, black, fur-covered figure "walking like a toddler," then falling and getting up again in berry bushes by beach just south of Metlakatla, Annette Island. [map5]

[T80A] 1980: Thom's Creek, south Wrangell Island, Chuck Trayler and Stan Hewitt fly in circles over long chain of biped tracks in snow on mountain. [map3]

[S81A] 1981 (approx.), summer: second-hand report from a woman, K. B., who stated her mother had told her of watching a large, black, hairy manlike creature fifty feet away from their fishing boat, just onshore at Mink Bay, extreme southeast Alaska. [map3]

[S81B] 1981, summer: Joe B. reported seeing a tall dark sasquatch watching his boat from shore near Cowboy Camp, a deserted village site north of Metlakatla on Annette Island. [map5]

[S82B] 1982: Ketchikan Lakes., Revilla' Is., AK, James Edenshaw reports watching a ten-foot-tall, dark sasquatch sleeping or resting in cave. [map5]

[S82C] 1982 or 1983: Last Chance Campground, Revilla Is., Mr. S. S. reported a seven- to eight-foot sasquatch standing in creek at night. [map3]

[V82A] 1982 or 1983: Dall Head, Gravina Is., AK, N. J. reported long scream "louder than human." [map5]

[S84A] 1984: west of Thorne Bay, P.O.W. Is., Ms. L. A. reports a seven-foot sasquatch running 35 mph beside car. [map4]

[T84A] 1984 (approx.), Nov.: Labouchere Bay, northwest P.O.W. Is.,

Bruce Shirley and son follow two sets of tracks over mountain, one set eighteen inches long. [map2]

[N84A] 1984 (approx.): Mahoney Mine, Revilla' Is., Harold Omacht report a nest in disused mine, nest "appeared made by hands." [map5]

[S85A] 1985, July: Soapstone Bay, North Yacobi Island, fisherman Tom J. Fisher and crew report watching a large brown bear become alarmed at something and run off, followed by a nine-foot black sasquatch stepping out on beach, watching the boat and then quickly running away from the bear's last position. [map1]

[T85A] 1985 (approx.) Nov.: Flicker Creek, north shore P.O.W. Is., three men report fifteen-inch tracks in snow at end of logging road. [map2]

[T85B] 1985 (approx.): Patrick Lake, Wrangell Is., Harvey Gross reports foot-long biped tracks in fresh snow on remote road. [map3]

[V86A] 1986: Lake Creek, Unuk R., AK, B. P. and two men reported three loud calls, "like a monkey whoop." [map3]

[S86C] 1986: Bugge's Beach, Saxman, Revilla' Is., two boys sight tall, dark sasquatch walking trail by ocean. [map5]

[V86B] 1986 (approx.): near head of Smeaton Bay, AK, hunter M. M. reported loud scream, not like a cougar. [map3]

[T86A] 1986 late Oct.: quarter-mile northwest of Kasaan, east P.O.W. Is., Robert Durgan described sixteen-inch tracks seen in snow while hunting, agreed with father and uncle that tracks did not resemble a bear or typical human. [map4]

[B87A] 1987 or 1988: north of Klawock, P.O.W. Is., two men with deer on beach see twelve-inch rocks thrown at them. [map4]

[NN87] 1987 (approx.): Tsimshean hunter described having pine-cones thrown at his back during evening, while dropped off alone deer hunting at remote location on Duke Island, south of Annette Island. He ascribed the behavior to a *ba'oosh*. [map3]

[S87A] 1987: Labouchere Bay, north P.O.W. Island, Brian E. reports seeing a five-foot black creature running along top of clearcut. [map2]

[B87B] 1987–89: Ward Lake, Revilla' Is., Brenda R., second-hand report of sighting, also "neighbors' car scratched by big creature walking on two legs." [map5]

[B88A] 1988: Affleck Channel, southern Kuiu Is., AK, Klawock fishermen V. S. and W. D. report heavy running footsteps rocking big boat at night; it dove off into ocean. [map2]

[N88A] 1988: Harriet Hunt-Leask divide, Revilla' Is., Jim L. reported fresh four-foot diameter nest, early fall, hair over four inches long

found. [map5]
[N88B] 1987: east of Klawock Lk., P.O.W. Is., logging engineer reports an eight-foot diameter woven cedar nest with long hair in it. [map4]
[V88A] 1988: Shinaku Inlet, west side P.O.W. Is., L. J. and D. L. report loud scream "Waaughh," "not a bear," witness stated. [map4]
[B89A] 1989: Gem Cove, Revilla' Is., Mr. S. B. and E. K. report theft of deer carcass in muskeg within minutes of killing deer with high-caliber rifle. [map5]
[S90A] 1990: south of Petersburg, Peter Byrne reports Tony Hawkins saw sasquatch head and shoulders in sea from boat. [map3]
[S90B] early 1990s: Prince William Sound, AK, Dr. Henner Fahrenbach reports Ed L. sighting a sasquatch with long mane of hair. [map1]
[S91A] 1991: south of Sunny Cove, northeast Cleveland Peninsula, M. M., taxidermist, reports scoping seven-foot pregnant female sasquatch on beach, later fifteen-inch tracks in gravel. [map3]
[B91A] 1991: Connell Road, Revilla Is., AK, Larry Thomas report, second-hand account from friend alleged knocked out by rock thrown from bush, same spot as Thomas' sighting in 1992. [map5]
[N91A] 1991: Second Waterfall Creek, Revilla' Is., military officer reported a twelve-foot diameter nest with logs four feet long, four inches thick laid in. [map5]
[S92A] 1992: Connell Road, Revilla' Is., AK, Larry Thomas reports seeing a seven-foot sasquatch with white eye-shine. [map5]
[V92A] 1992: Lake St. Nicholas Rd., west P.O.W. Is., Bruce C. with brother reports several loud screams, "not like a bear." They leave gear, spend night elsewhere, return next day. [map4]
[S92B] 1992: Crab Bay, Annette Is., AK, fisherman Dan A., with father, reports beige man-sized sasquatch on beach. [map4]
[S92C] 1992: Westshore of Ernest Sound, east of Wrangell, Ketchikan man reports watching pregnant female on beach through high-power riflescope.
[S92D] 1992: Work Channel, Alaska/B.C. border, Peter Byrne related Eric Sowers report a sasquatch swimming near boat, then diving. [map7]
[S93A] 1993: south of Klawock Lake., P.O.W. Is., AK, Mr. I. W. and wife reported tall, dark sasquatch crossing highway. [map4]
[V93A] 1993: Old Frank's Divide, central P.O.W. Is., forestry professional D. H. reports loud calls, "somewhat like a monkey." [map4]
[S93B] 1993: above Carlanna Lake, Revilla' Is., Mr. M. A. reported sighting black, seven-foot sasquatch near Ketchikan, later found one fifteen-inch track. [map5]
[S94A] 1994: north of Saxman, Revilla' Island, AK, D. P. with friend

reported eight- to ten-foot dark sasquatch peeling bark. [map5]
[V94A] 1994: Eek Mt. near Hydaburg, P.O.W. Is., Rob Sanderson reported four loud whistles and wooden rapping noise. [map4]
[T94A] 1994–97: above Yes Bay, AK, Mr. S. B. reported family finding large tracks up old roads. [map3]
[S95A] 1995: southwest of Klawock Lk., P.O.W. Is., AK, forestry professional reported second-hand sighting of dark brown sasquatch by girl and boyfriend. [map4]
[B95A] 1995 (approx.): north of Thorne Bay, P.O.W. Is., hunter Victor Mulder, with two others, reported strange uniform piles of deer bones with no marrow, no saw marks on broken long bones, not scattered by scavengers. [map4]
[S96A] 1996: Bear Mt., Ketchikan, AK, Mr. S. B. reported shooting at "bear." When it "got up on hind legs," after his shot, it walked off with "arms swinging," hunter reported he dropped his rifle. [map5]
[V96A] 1996: Bostwick Inlet, Gravina Is., AK, Mr. L. D. and uncle reported a scream up creek, camp found vandalized on their return. [map5]
[T97A] 1997: north of Naukati, north P.O.W. Is., deer hunters J. C. and T. A. reported finding over 100 biped tracks in snow; tracks fifteen and a half inches long. [map3]
[S97A] 1997: Walker Channel, forty miles south of Sitka, west coast of Baranof Island, fisherman H. A. and others saw an eight-to nine-foot, dark brown sasquatch. [map2]
[V97A] 1997: Tombstone Bay, Portland Canal, AK, sole inhabitant John Kristovich reported long yell in daytime from forest several hundred yards away, louder than bear or human. [map6]
[S97B] 1997: Hidden Glacier, north of Hyder, AK, Frank S. with snowmobilers reported a big sasquatch on snowfield at night, it screamed and walked toward them. [map6]
[S97C] 1997 or 1998, summer: Mr. S. W. and son report a nine- to ten-foot-tall dark sasquatch crossing highway in vicinity of south end of Klawock Lake, P.O.W. Is., at 10:00 p.m. [map4]
[S97D] 1997, July: Stanley Edenshaw and step-brother Mickey Calhoun, driving to Hydaburg at 2:30 a.m., P.O.W. Is., saw a seven- or eight-foot, dark brown sasquatch walk quickly across the Hydaburg highway and into the forest. [map4]
[S98B] 1998, March 1: Ketchikan family sees seven foot, dark brown sasquatch walk, then run across highway at Whipple Creek Bridge. [map5]
[S98A] 1998, summer: mile 12, north of Ketchikan, Revilla Is., cyclist

Gerald P. reported passing brown sasquatch digging in ditch beside highway. [map5]

[T98A] 1998: Whipple Spur, Revilla Is., land manager E. E. reported a fifteen-inch track by muskeg, also deer tracks. [map5]

[V98A] 1998: east side Betton Is., west of Revilla' Is., west Behm Canal, Rob Shelton with friends reported two loud screams from forest on Betton Island, at night. [map5]

[B98A] 1998: east side Thorne Arm, Revilla' Is., miner Jake Lauth reported lid ripped from dynamite freezer with no claw scratches or pry marks, nothing taken. [map5]

[B98B] 1998: Upper Wolf Lake, West Revilla Island, A.K., two hunters searching for black bear wounded the previous night find bear carcass stuffed in muskeg hole and covered with patchwork of sphagnum moss. Hunters state no one else in area, and that moss camouflage required hands to do. [map5]

[V98B] 1998: Rudyerd Bay, east Behm Canal, Ms. S. L. reported long series of angry yells from shore in remote setting. [map3]

[T98A] 1998: Little Totem Bay, south tip of Kupreanof Is. near Wrangell, trapper reported barefoot "size 10" tracks in sand. [map2]

[V98B] 1998: Gravellies Ck., Thorne Bay, east P.O.W. Is., Ms. S. B. with boyfriend reported hearing scream and "wood-on-wood" sound. [map4]

[V98C] 1998: west side Thorne Arm, Revilla' Is., S. L. reported scream "louder than bear could make." [map5]

[V98D] 1998: east side Thorne Arm, Revilla' Is., prospector Jake Lauth reported "rocks being clacked together" and logs breaking. [map5]

[V98E] 1998: end of Brown Mt. Road, Revilla' Is., hunter M. D. reported scream "louder than jet engine" from trees. [map5]

[V98F] 1998: end of Brown Mt. Road, Revilla' Is., Ray S., second-hand report, friends reported loud rapping, trees breaking, rhythmic "booming" on slope above road, close to previous report. [map5]

[V98G] 1999: Gertrude Lake, Green Monster Mt., central P.O.W. Is., AK, hunter Bob E. with companions reported a strange, ten-second-long call, "louder than elk." [map4]

[T99A] 1999: north of Boca de Quadra Inlet, AK, Frank S. and girlfriend report fresh string of biped fifteen-inch, barefoot tracks on beach, with twelve-foot stride between each left footprint. [map3]

[S99A] 1999: Saxman, AK, Mike V. and others reported a seven-foot, dark sasquatch cross service road in front of parked car past midnight. [map5]

[S99B] 1999, winter: Ward Lake, Revilla' Is., AK, anonymous Ketchikan witness reported a six-foot light brown sasquatch standing in ditch at 5:00 a.m. as she drove by it. Sasquatch was pulling at old salmonberry branches. [map5]

[N99A] 1999: east of Klawock Lake, P.O.W. Is., Al Jackson and author photograph three inverted cedars jammed in muskeg up logging road, no logging marks, rock cairns were found nearby. [map4]

[MB99] 1999, fall: Connell Creek, Revilla' Is., two men independently reported finding remains of an eight-inch hairy foot in sand by creek, not belonging to a bear or man. Both men respectively alleged they each threw it back. [map5]

[T00A] 2000, January: mainland opposite Reef Is., Portland Canal, AK, fishermen Steve S. and Jeff J. reported string of biped tracks in snow four feet deep, tracks coming down from forest into water. [map6]

[S00A] 2000, February: Kake, Kupreanof Is., AK, Mr. H. K. with girlfriend reported six-foot dark brown or black sasquatch leaping ditch in front of car at night. [map2]

[S00B] 2000, June: George Inlet, Revilla' Is., James Nunez reported seeing a seven- to eight-foot sasquatch at 100 yards through a nine-power scope in early morning. It had been sitting on a rock watching him but backed away while he held his rifle on it. A sixteen-inch track was later found. [map5]

[S00C] 2000, early August: Hydaburg, P.O.W. Is., Stan Edenshaw reported seeing a dark brown, seven-foot "Hairy Man" walking across the old gravel quarry one mile out of Hydaburg, 9:30 at night. [map4]

[S00D] 2000, October: a couple driving between the Hydaburg cutoff and Hollis at 8:30 p.m., saw a dark, hair-covered creature taller than a man walk quickly across the highway. "It wasn't a bear, I know what I saw," stated wife. [map4]

[B00A] 2000: south of Ketchikan, Revilla' Island, AK, Mrs. S. H. and husband reported 600-pound furnace outside house picked up and thrown overnight. [map5]

[V00A] 2000, June: Brown Mountain Road, southwest Revilla' Island, AK, Dave Timmerman, Ketchikan, reports hearing a yelling call louder than human or wolf on Brown Mtn., also noted large tree with a horizontal limb debarked in a manner atypical of bears, no sign of human involvement. [map5]

[S01A] 2001, Jan. 2: Klawock-Hollis Highway, south of Harris River, central P.O.W. Is., AK, husband and wife reported seeing eight-foot dark brown hair-covered creature, "definitely not a bear or human," walk across highway at 5:00 a.m., creature crouched

	down as car slows and approaches, eye-shine red. [map4]
[S01B]	2001, June 8: south of Ketchikan, Revilla' Island, AK, Mr. D. Johnson stated that at 9:30 p.m., driving two miles north of Herring Bay, he saw a man-sized creature, coated with dark, dripping, wet hair, walk on two legs across the road forty to sixty yards in front of his car, crossing from oceanside to steep inland side. [map5]
[S01C]	2001, July: Montana Creek Trail, west of Juneau, AK, Ms. S. W. reported that she and two friends watched a six and a half foot dark brown sasquatch with reddish eye-shine approach their bonfire to within a distance of fifteen feet. [map2]
[S01D]	2001, July: Herring Cove, Revilla' Is., AK, witness reported seeing a six-foot dark brown, hair-covered figure running "faster than a bear" along the south shore of Herring Cove at 9:30 p.m. [map5]
[S01E]	2001, July: Herring Cove, Revilla Is., AK, second-hand report of anonymous man seeing a man-sized, brown sasquatch running very fast across a meadow south of Herring Creek, several hundred yds. inland from previous report. [map5]
[S01F]	2001, April: Last Chance Campground, Revilla Island, AK, second-hand report from a military officer who told friend he had seen a sasquatch, seven feet tall or more, watching him at night from twenty feet, then cross highway. [map5]
[V01A]	2001, June: Exchange Cove, northeast P.O.W. Is., two men camped in tent hear unusually loud chattering, chuckling sound during night, nothing seen next morning. [map2]
[V01B]	2001, late July: Carlanna Lake, Revilla Island, five youths camping alone report being awakened at 2:30 a.m. by the loud sound of "a baby crying from the forest." [map5]
[N01A]	2001, early August: Carlanna Lake, Revilla Island: Author and B. Johnstone Sr. examined a forty-foot, four-inch diameter green alder freshly snapped off at four feet above the ground, bent over horizontally with no marks. "Not man or bear, " observed Johnstone. [map5]
[T02B]	2002, February 2: S. Point Higgins Road, Revilla' Is., Andy Rasmussen reported biped tracks in snow entering forest behind customers house on a fuel delivery in evening. Witness stated they had a stride longer than a human and showed toe marks and impressions of fur around edges in 6 inch snow. [map5]
[T02C]	2002, spring: Jody Mazzama, Ketchikan, reported several curious fifteen-inch five-toed human-like tracks at the edge of Herring Cove, in mud and snow, beside the Herring Cove bridge. [map5]

[T02D] 2002, July: White River, Revilla' Is., Mike V. and Emily G., with friend, beachcombing a small island near river mouth, reported a long string of two sets of five-toed tracks, approx. twenty inches and sixteen inches, the larger with a five-foot stride, in sand on beach. [map5]

[S02A] 2002, mid-August: Hydaburg, Prince of Wales Island, investigator Al Jackson reported that Bert C. and wife, Klawock, were driving north out of Hydaburg, around 11:30 p.m., and sighted a dark hair-covered sasquatch walking across highway six miles south of Klawock. Creature said to have long hair on back of head and neck flowing down back of shoulders, "kind of like a horses' mane." [map4]

[S02B] 2002, mid-August: Sitka, Baranof Island. Two local women reported an evening sighting of a tall, dark-haired sasquatch walking into the forest along the old pulp mill road.

[S02C] 2002, mid-October, two miles north of Craig, Prince of Wales Island. A Ketchikan woman and a Craig taxi driver reported seeing a seven-foot, dark-haired sasquatch run across the highway at 1:00 a.m. The witness said the sasquatch ran into a small gravel quarry. The driver refused the woman's request to turn off to follow it.

Reports in southeast Alaska are rarely passed on except by word-of-mouth. In addition, it is a conservative estimate that there are as many reports for each of the five following population areas as for the Ketchikan area: Craig (including Hydaburg and south Prince of Wales), Thorne Bay (including Klawock, Coffman Cove, Naukati and north Prince of Wales), Sitka (including Baranof Island), Wrangell (including Petersburg area and Kake), and Juneau (including Pelican, Haines, Skagway, Admiralty Island, Hoonah and the rest of the northern part of Southeast Alaska to Yakutat).

The few reports listed here from these other areas have mostly been obtained through individuals who have relocated to the Ketchikan research area. It is estimated that up to several hundred more reports of all preceding types might be assembled through local research in these other regions of southeast Alaska. In this sense, what is presented here may provide no more than a local (Ketchikan area) sampling of many similar sasquatch sightings elsewhere in southeast Alaska.

There is also the problem, already mentioned, of sightings not passed on at all. After ten years of travel, interviewing and investigation in southeast Alaska, it appears that fewer than 10 percent of all reports in this area are passed on beyond immediate family or close friends. The actual number of sightings from this area of Alaska may never be known. The following appendix for reports from British Columbia indicate that, so far, coastal Canadians may be slightly ahead in the above categories, possibly having fewer inhibitions about mentioning sasquatches or perhaps simply more forests with sasquatches in them.

Appendix Three

A Summary List of B.C. Coastal Reports

The following is a chronological listing of documented sasquatch sighting and track reports from coastal British Columbia. Some reports from inland regions are included where they occur within 100 miles of tidewater.

Reference key

NAGATS: ......*North America's Great Ape: The Sasquatch*, John Bindernagel.

RA: ....*Rob Alley, personal files.

SAS: .... *Sasquatch*, Rene Dahinden.

SBTCM: ....*Sasquatch/Bigfoot: The Continuing Mystery*, Thomas Steenburg.

STAAU: ....*Sasquatch: The Apes Among Us*, John Green.

TBC: ....*The Bigfoot Casebook*, Colin and Janet Bord.

TSF: ....*The Sasquatch File*, John Green.

TTR: ....*The Track Record*, newsletter of The Western Bigfoot Society, ed. Ray Crowe, Hillsboro, OR.

<www.bfro.net>: The web site of The Bigfoot Researchers' Organization, host: Matt Moneymaker.

(OS84) 1884, late June: newspaper account of capture of "Jacko," 4'7" tall, 127 lb., 1" hair, near Yale. (TSF: 5) (map12)
(OS00) 1900 (approx.): man reported very large, barefoot, humanlike tracks in sand beside Harrison River. (TSF: 9) (map12)
(S04A) 1904, December: newspaper account of young hair-covered wild man seen by four men near Horne Lake, Vancouver Island. (TSF: 9) (map11)
(S05A) 1905, April: Two old Victoria *Colonist* accounts, April, 1905, and May 2, 1905, of "local Indians between Union Bay and Comox firing at what they took for a bear digging on a beach. It straightened up, yelled, and ran upright into the woods. Also noted it had been digging clams. (map11)
(S07A) 1907, March: newspaper account of "monkey-like wild man" digging clams, howling at Bishop's Cove, causing desertion of village, presumably the former Native village northeast of Bishop's Bay, west of Kemano on the Gardner Canal. (TSF: 9) (NAGATS: 86) (map8)
(S09A) 1909 (approx.),May: P. Williams reported that a six and a half foot sasquatch chased him near Chehalis Reserve near Agassiz. (TSF: 10) (map12)
(S00A) early 1900s: Victoria *Colonist*, Dec. 15, 1904, and Vancouver *Province*, Dec. 21 and 30, 1904, carried reports of famous timber cruiser Mike King sighting "monkey-man," reddish brown, long arms; washing piles of wild onions or grass roots in mountains near Campbell River, Vancouver Island. (TSF: 9) (NAGATS: 212) (map11)
(S00B) early 1900s: Kitimaat Village elder recounted shooting a sasquatch, other creatures chased him. (TSF: 9) (map8)
(S15A) 1915: Charles Flood swore affidavit of sighting eight-foot, light brown sasquatch, twenty-five miles south of Hope in the Fraser Valley. (TSF: 12) (map 12)
(S24A) 1924, summer: Albert Ostman swore report of abduction by eight-foot male sasquatch near Toba Inlet on the central coast, reported held captive with family of four sasquatches for several days in high rock-enclosed meadow. (TSF: 12) (map11)
(S27A) 1927, late September: Native hop-pickers report close sighting of sasquatch on railroad near Agassiz in Fraser Valley. (TSF: 11) (map12)
(S28A) 1928: a Bella Coola trapper, Mr. George Talleo, shot at a sasquatch which fell, got up, then ran. East of South Bentick Arm, central coast. (TSF: 14) In other details from this sighting, Mr.Talleo reported finding "moss-covered excrement" shortly

(S30A) before. (NAGATS: 213) (map10)
1930s: Ellen Neal, renowned Native carver, reported that as a child, saw hirsute giant on beach, Turnour Island near Alert Bay; tracks found later. (TSF: 14) (map9)

(S30B) 1930 (approx.): Tom Brown, Klemtu, in a group at Kwakwa, Swindle Island, saw ape in ocean, it was shot at. (TSF: 14) (map8)

(S33A) 1933 (approx.): report of two Vancouver men seeing black sasquatch eating berries at a small high altitude lake above Pitt Lake, southern Coast Mountains. (TSF: 14) (map12)

(S39A) 1939: woman and two children saw sasquatch approaching their home near Ruby Creek, Upper Fraser Valley. They fled, sixteen-inch tracks found later. (TSF: 14) (map12)

(S40A) 1940s: Bella Coola bear guide Clayton Mack watched tall upright hair-covered creature walk over logs at Jacobsen Bay, near Bella Coola, central coast. (TSF: 14) (map10)

(S40B) mid-1940s: seiner skipper saw large erect hairy creature, dark brown, running at Jacobsen Bay, Bella Coola area. (TSF: 14) (map10)

(S40C) early 1940s: Alex Oakes, saw seven-foot brown sasquatch with long hair running across road one mile east of Coombes, central Vancouver Island. (TSF: 14) (map11)

(T40A) 1940s: farmer reported "huge barefoot tracks" in snow leading from a cow field onto a mountain slope near Agassiz, Fraser Valley. (TSF: 14) (map12)

(T45A) 1945, April: Tom Brown reported huge tracks near a beach named in Native language "where the apes stay," Swindle Island. (TSF: 14) (map8)

(T47A) 1947 (approx.): a Fort Rupert woman reported to John Bindernagel a set of tracks she saw on a Turnour Island beach about 1947. They were barefoot and manlike, but larger and the big toe imprint showed great flexibility, they "stuck out to the side at a 45-degree angle." (NAGATS: 55, 213) (map9)

(S48A) 1948 (approx.): H. Charlie reported sasquatch chased him on his bicycle one mile on Morris Valley Road, near Harrison Mills. (TSF: 14) (map12)

(S49A) 1949, October: a highway foreman reported an eight-foot dark brown heavy animal on road south of New Hazelton in Skeena Valley. (TSF: 14) (map 8)

(T50A) 1950 (approx.), winter: Paul Hopkins reported large humanlike tracks on shore at Jackson Passage, Roderick Island. (TSF: 14) (map8)

(T50B) 1950 or 1951: two men reported humanlike tracks of different sizes, some fourteen inches long, cross a sandbar and lead into Kitasu Lake, Swindle Island. (TSF: 18) (map8)
(S53A) 1953, September: Jack Twist saw a dark, eight-foot-tall hair covered creature walk off a logging road near Courtenay, Vancouver Island. (TSF: 18) (map11)
(S56A) 1956, May: south Hunt saw grey-haired, seven-foot upright creature cross highway near Hope, glimpsed one more. (TSF: 18) (map12)
(S56A) 1956, (approx.), August: a Port Hardy man related to J. Bindernagel a report from three Bella Coola moose hunters who watched a sasquatch eating blueberries inland from Bella Coola in August around 1956. It had a humanlike face, "except for its protruding mouth." (NAGATS, pp. 38, 210.) (map 10)
(S50A) late 1950s: man reported a huge, dark brown, erect, hairy "beast" cross the road thirteen miles south of Hazelton. (TSF: 18) (map8)
(S58A) 1958, December: two hunters reported a seven-foot tall, black sasquatch watch them, then run into timber, thirty miles east of Bella Coola. (TSF: 18) (map10)
(S59A) 1959, March: Laurence Hopkins reported an "ape" emerge from the bush just behind him on the beach, he ran and swam to avoid it, Aristazabal Island, northern B.C. coast. (TSF: 18) (map8)
(S60A) 1960, February: a man stated he saw an erect, apelike creature, "the size of a small man, " Price Island, northern B.C. coast. (TSF: 18) (map8)
(S60B) late 1960s: a caver reported a seven-foot-tall sasquatch while exiting a cave at 3:00 a.m., central Vancouver Island. (TSF: 41) (map11)
(S60C) late 1960s: a Victoria man provided author with a second-hand report of a military officer, based at CFB Holberg, extreme northern tip of Vancouver Island, who had been driving his wife and father-in-law home along the Holberg Road. The officer noticed what he thought was a huge, dark man lying face-down in the ditch as they drove by. He wanted to stop and offer help but his Native father-in-law persuaded him otherwise, saying "You just wouldn't understand these things." Returning on the road later that trip, officer stopped, noted an eight-foot area in ditch where the grass had been pressed flat. (RA files) (map9)
(S60D) late 1960s: an Alert Bay resident described to John Bindernagel an event on Turnour Island where two adults with a group of schoolchildren "were awakened in the night by a strong foul smell followed by forceful striking of the house posts which

shook the building." Next morning a bag of clams was missing from porch. (NAGATS: 112) (map9)

(B60A) 1960, August: a woman and family tenting in remote campsite at Cultus Lake, upper Fraser Valley, experienced forty to fifty stones up to one and a half inches diameter thrown at tent for more than two hours at night. In morning a new trail found made in bushes nearby. (NAGATS: 117) (map12)

(S60B) 1960 (approx.), winter: two men stated they "shot at a small ape" at Watson Bay, Roderick Island. (TSF: 18) (map8)

(T61A) 1961, April: John Green investigated a reported track on the south bank of the Fraser River, ten inches wide, twelve inches long, with five toes, near Agassiz. (TSF: 25) (map12)

(T61B) 1961, July: John Green investigated a set of wide "size 11" tracks on a beach at Swindle Island. (TSF: 25) (map8)

(T61C) 1961, October: Bob Titmus investigated fresh thirteen and a half inch tracks, four feet apart, exiting the ocean near Swindle Island. (TSF: 25) (map8)

(T61D) 1961, December: Bob Titmus found sasquatch tracks in snow made overnight at an abandoned cannery at Klemtu. (TSF: 26) (map8)

(S62A) 1962, April: a woman stated she saw a female sasquatch holding a smaller sasquatch by the hand on the riverbank near town, Bella Coola River. (TSF: 26) (map10)

(T62A) 1962, summer: Bob Titmus found 1,200 yards of biped tracks larger than a man's on Aristazabal Island. (TSF: 26) (map8)

(T62B) 1962, August: Bob Titmus found fourteen-inch, five-toed tracks, forty-two inches apart by a creek on a small island in Devastation Channel. (TSF: 26) (map8)

(T62C) 1962, fall: Bob Titmus found three sets of tracks, approximately 14, 13 and 12 inches, on a beach on Swindle Island. (TSF: 26) (map8)

(T62D) 1962, December: Bob Titmus found four sets of adult sasquatch tracks in deep snow in Tweedsmuir Park, central Coast Mountains. (TSF: 26) (map10)

(N63A) 1963: Ray Roberts, beachcombing Price Island, midcoast, reported rotting smell, investigated and found eight-foot diameter nest of grass and other vegetation covered with crude but substantial shelter of poles and logs. The low entrance faced the forest. (NAGATS: 69, 214) (map8)

(S63A) 1963: a man reported to John Green of seeing a large creature rise up and walk on the tide flats at the head of Thompson Sound. (TSF: 26) (map9)

(S63B) 1963: J. Wilson reported to Bob Titmus of seeing a sasquatch on the shore of an island near Bella Bella, central coast. (TSF: 27) (map9)
(S63C) 1963: Bob Titmus observed three brown figures climb a cliff using arm and leg movements like humans, across a valley south of Kemano, north B.C. coast. (TSF: 27) (map8)
(B64A) 1964, April: J. Green checked an Ellen Neal report of a house damaged by a sasquatch at Turnour Island, central B.C. coast. The house was damaged somewhat, the man rescinded his report to "a bear" but immediately clearcut all his surrounding property of dense timber. (TSF: 27) (map9)
(S64A) 1964, July 4: two youths reported they saw two seven-foot, shaggy, dark creatures at 2:30 a.m. near Chilliwack, Fraser Valley. (TSF: 27) (map12)
(T64A) 1964 (approx.): Mrs. J. Robertson reported fourteen by seven inch short-toed tracks by the Fraser River near Agassiz, Fraser Valley. (TSF: 27) (map12)
(T64B) 1964 (approx.): a man reported to Rene Dahinden a trail of nineteen-inch biped tracks in snow n. of Squamish, south B.C. mainland coast. (TSF: 27) (map12)
(T64C) 1964 or 1965, summer: a Klemtu man reported yards of wide, biped fourteen-inch tracks near Meyers Pass. (TSF: 33) (map8)
(S65A) 1965, May 31: Mrs. Seraphine Jasper reported a black, tall sasquatch in a field on Nicomen Island near Mission, Fraser Valley. (TSF: 33) (map12)
(ST65) 1965, June: a man stated he saw tracks that measured twenty-four inches long and then saw a twelve-foot-tall auburn sasquatch, at a frozen lake northwest of Pitt Lake, lower Coast Mountains. (TSF: 33) (map12)
(S65B) 1965, July: Jack Taylor, Butedale, reported two huge bipeds at seventy-five yards, one swimming, opposite Butedale. (TSF: 33) "The one in the water swam very powerfully and very fast, with the water surging around its chest." (NAGATS: 44, 212) (map8)
(S65C) 1965, summer: a group saw a huge, dark brown sasquatch on a small rock offshore, near Clio Bay near Kitimat, north B.C. coast. (TSF: 33) (map8)
(S65D) 1965, fall: a man reported a small, dark female creature with long hair in the bush just off the road near Harrison Mills, Fraser Valley. (TSF: 33) (map12)
(S66A) 1966, July: two Gilmore brothers reported a dark, biped animal in a pine bog adjacent their farm running cattle on Lulu Island, Fraser Valley. (TSF: 33) (map12)

(S66B) 1966, July 21: A woman reported seeing the head and shoulders of a manlike thing over six feet tall in salmonberry bushes beside her house near Lulu Island, Fraser Valley. (TSF: 33) (map12)
(S66C) 1966, July 22: John McKernan reported seeing an upright, hairy giant cross a road into bushes by Fraser River. (TSF: 33) (map12)
(S66D) 1966, July: Mr. LeToul reported to RCMP a sasquatch passed by his house near White Rock, lower B.C. mainland. (TSF: 34) (map12)
(T66A) 1966, November: two hunters reported strange thirteen-inch tracks in snow circling them, thirty miles east of Bella Coola, central coast. (TSF: 34) (map10)
(T67A) 1967, February: Guide C. Mack reported "ape" tracks, also sideways on logs, two miles up Quatna River, central coast. (TSF: 34) (map9)
(S67A) 1967, September: two teenage girls on cobble beach on east shore of Vancouver Island encounter a dripping wet, seven-foot reddish brown sasquatch holding a stick in one hand and three or four ducks in the other. After several seconds both sasquatch and girls fled in opposite directions. (NAGATS: 96, 97, 126, sketch 97) [Sketch by one of the eyewitnesses appears in Fig. 54.]
(SV67) 1967, February: two men reported an ape on an island that they shot at. "It screamed like a woman," Hartley Bay. (TSF: 34) (map8)
(T67B) 1967–68, winter: Paul Hopkins told Bob Titmus he saw thirteen- to fourteen-inch ape tracks near a beach on Swindle Island. (TSF: 34) (map8)
(S68A) 1968, February: newspaper report of two men seeing a six-foot-tall sasquatch on a clam beach at dusk on Broughton Island, central B.C. coast. (TSF: 41) (map10)
(ST68) 1968, May: Gordon Baum reported watching a five-foot-tall, black sasquatch, with a track six inches wide, on a logging sidehill near Sechelt, south B.C. coast. (TSF: 41) (map11)
(S68B) 1968, mid-August: two men reported a seven-foot or taller sasquatch with no neck and a flat nose at 4,000 feet in elevation, near Stewart, north B.C. coast. (TSF: 41) (map6)
(S68C) 1968 (approx.): a Victoria man reported to author that he, his uncle and several cousins had been beach hiking early a.m. at Sombrio Beach, between Sooke and Port Renfrew, southwest Vancouver Island. As fog lifted, they all noticed a seven-foot, dark, heavy set manlike figure step over a large log and walk up beach into the forest. Returning later they noticed large tracks by the log. (RA files) (map11)

(S68E) 1968 or 1969, October or November: a B.C. forestry worker near Gordon River, southwest Vancouver Island, reported to John Bindernagel that a hunter had found two buck deer heads, apparently "twisted off, " numerous sasquatch tracks in area. (NAGATS: 73, 216) (map11)
(S69A) 1969, February: Ronnie Nyce reported that he and two other men fired at a sasquatch which ran off, Khutze Inlet. (TSF: 41) (map11)
(T69A) 1969, March: a man reported that a cabin owner at a lake at 6,000-foot elevation saw barefoot tracks with a six-foot stride at Lewis Lake, Cheakamus Valley, lower B.C. Coast Mountains. (TSF: 41) (map12)
(T69B) 1969, April: workmen reported several miles of fourteen-inch tracks in snow and a trail of spruce buds in tracks between clumps of trees above Callaghan Valley, lower B.C. Coast Mountains. (TSF: 41) (map12)
(T69C) 1969, June 18: Sam Brown reported he and two others saw fifteen-inch tracks, four and a half feet apart, at Kitasu Lake, Klemtu, north B.C. coast. (TSF: 41) (map8)
(S69A) 1969: Bob Titmus received reports of two sasquatch sightings near Massett, Queen Charlotte Islands. (TSF: 41) (map7)
(SB70) 1970, January: road foreman Bill Taylor reported a seven-foot-tall sasquatch run across the road, fish in one hand, twenty miles north of Squamish, lower B.C. coast. (TSF: 51) (map12)
(S70A) 1970, February: Andrew Robinson reported seeing a six-foot-tall, light brown-colored creature walk on a beach one mile south of Klemtu, north B.C. coast. (TSF: 51) (map8)
(S70B) 1970, March: Tina Brown reported a seven-foot "more like a gorilla than a man" at twenty feet in her headlights, north of Skidegate, Haida Gwaii (Queen Charlotte Is.). (TSF: 51) (map7)
(S70C) early 1970: a boy reported seeing a creature similar to the above, run off with its arms lifted, near Juskatla, Haida Gwaii (Queen Charlotte Is.). (TSF: 51) (map7)
(S70D) 1970, April: a man reported nearly hitting an eight-foot "man or gorilla" in midhighway near Harrison Hot Springs, Fraser Valley. (TSF: 51) (map12)
(S70E) 1970, April 23: the crew of the seiner *Bruce I* reported a big, beige, erect creature with an apelike face on the shore one mile north of Klemtu. (TSF: 51) (map8)
(S70F) mid-1970s: at Wakeman Sound on Kingcome Inlet, north of Broughton Island on the midcoast, a prospector reported to John Bindernagel that he had found a nest up in the center of a clump

of mountain hemlock. "branches had been woven in and out around the outside rim of the nest." (NAGATS: 69, 214) (map9)

(T71A) 1971, January: two researchers checking a report found prints and seven inch or longer reddish brown hairs on Graham Island, Queen Charlotte Islands. (TSF: 51) (map7)

(T71B) 1971, winter: Mrs. E. Yerxa reported miles of fourteen-inch tracks with a thirty-to thirty-six-inch stride, fourteen miles east of Hope, Fraser Valley. (TSF: 51) (map12)

(T71C) 1971, May: G. Conway reported fourteen-inch tracks in sand, thirty-plus miles up a logging road in the Squamish Valley, southern B.C. Coast Mountains. (TSF: 51) (map12)

(S71A) 1971, October: a man reported two large human-shaped, beige creatures pulling lily roots at a lake north of Sechelt, south B.C. coast. (TSF: 51) (map11)

(S73A) 1973, March 21: Peter Spika, Luko Burmas and Nick Pisac, herring fishermen, watched a beige, ten-foot tall sasquatch walking on a beach in Bute Inlet, lower B.C. mainland. (Witnesses interviewed by John Green and Bob Titmus, noted by Green, "Not All Quiet on the Western Front," *Pursuit*, vol. 7, no. 4, p. 98) (map11)

(ST73) 1973, June: a timber worker saw a sasquatch jumping up and down leaving "humanlike" footprints near Sechelt. (*Sechelt-Peninsula Times* article, noted by Green, STAAU: 420) (map11)

(S73B) 1973, July: a fisheries patrolman watched a six-foot-tall gray sasquatch rooting in vegetation on a beach in daylight at Roscoe Inlet. (Interviewed by J. Green and Bob Titmus in article noted above, also noted TBC: 195) (map8)

(B73A) 1973, summer: strange sounds were reported heard at a summer camp; animals were reported missing near Langley on the north shore of the Fraser Valley. (www.bfro.net) (map12)

(SB73) 1973, fall: Clayton Mack and California hunter watched black sasquatch eating roots on beach at head of South Bentick Arm. (Excerpt from Mack's book, *Grizzlies and White Guys*, for reference see "Old Stories from the Coast.") (map10)

(T73A) 1973, winter: two men hiking through snow in dense timber on old logging road above Sooke, south Vancouver Island, report hearing crashing sounds in trees to one side, retracing their steps find seventeen-inch fresh tracks, only minutes old, men later cast tracks. (RA files) (map11)

(S74A) 1974, July: Wayne Jones reported to Rene Dahinden that he watched a seven- to eight-foot tall sasquatch from his campfire for five minutes, he stated it was licking mud off its fingers, near the head of Harrison Lake, southern Coast Mountains. (SAS:

(S74B) 196) (map12)
1974, summer: an elderly salal picker reported watching a thin, brown-colored sasquatch strip buds off a second-growth hemlock sapling while crouched ten feet away, near Qualicum, east central Vancouver Island. The man's girlfriend working nearby stated a feeling of being watched; their dog was also acting strangely. The man became afraid to return to work, asked RCMP for handgun ammunition for personal safety. (RA files) (map11)

(S75A) 1975, summer: Julius Szego and friend reported to Guy Phillips that he saw a sasquatch near Skutz Falls on the Cowichan River above Duncan, southeast Vancouver Island. (SBTCM: 68) (map11)

(B75A) 1975: Tom Sewid reported to John Bindernagel incident occurring to two boys playing on leaning totem at deserted village near Telegraph Cove, north Vancouver Island. Boys were subjected to a shower of small (less than one-inch diameter) stones thrown at them from forest. Boys decided to leave. (NAGATS: 116) (map9)

(S76A) 1976, January: a logging camp watchman reported sighting a sasquatch standing on the main logging road in snow just north of camp at the head of Knight Inlet. He became upset and was flown out for medical treatment. (RA files) (map10)

(T76A) 1976: in mud on banks of Skeena River near Terrace, Bob Titmus found and casted seventeen-inch tracks. (NAGATS: 58, photo) (map8)

(S77A) 1977, March 28: Richard Mitchell reported a seven and a half foot creature with a peculiar gait by the roadside near Spuzzum. (*Vancouver Sun* article, April 2, 1977) (map12)

(TB77) 1977: Bob Titmus, reinvestigating area of sasquatch tracks on Skeena River from previous year, found "many branches up to the size of my wrist were broken or twisted off about six feet above the ground in the adjacent bush and forest." (NAGATS: 79, 216) (map8)

(N77A) 1977 or 1979, summer: Professor Fred Bunnell reported to John Bindernagel that while studying grizzlies on the Ahtnuhati River, Knight Inlet, midcoast, he found a nest he designated as a possible sasquatch "bower, " made of bent, broken and over-arching branches piled up against a rock face, adding, "No bear makes a day bed like that." (NAGATS: 69, 214) (map10)

(S78A) 1978, September: Jesse Gordon, Alert Bay, reported encountering a sasquatch on a trail near Rock Bay, northeast Vancouver Island; its odor was "eye-watering." (NAGATS: 77, 217) (map11)

(V79A) late 1970s: John Bindernagel interviewed a man who had been living in a cabin near the Sooke River, southern Vancouver Island, who reported calls similar to Puyallup, Washington, purported sasquatch tape recordings, but "more melodious." (NAGATS: 123, 227) (map11)

(S80A) 1980, December 13: near Comox Lake, Vancouver Island, Ken Berkeley and hunting partner Terry Kerton reported sasquatch with long arm hair leaning against tree. At first they thought it "a guy in torn rain gear." (NAGATS: 12, 211) (map11)

(V82A) 1982: near Strathcona Park, Vancouver Island, wildlife biologist Dr. John Bindernagel documented apelike "whooping." (JB files) (map11)

(B82A) 1982 or 1983, summer: a man reported a youth with nineteen others at church camp being pelted with rocks during the night, glimpsed male figure running, on Gerald Island, Nanoose Bay, east central Vancouver Is. (© bfro.net) (map11)

(SV83) 1983, summer: hikers on Tetrahedron Ridge near Gibson's Landing, south B.C. mainland coast were followed by an eight- to ten-foot-tall black creature, later that night they hear a series of roars. (© bfro.net) (map11)

(S84A) 1984, summer: a resident of Victoria reported to John Bindernagel of seeing a sasquatch stand up in a patch of salal bush near a swimming hole by Englishman River, near Parksville, central Vancouver Island. (NAGATS: 208) (map11)

(S85A) 1985, September: near a roadside rest area, Nimpkish River, north Vancouver Island, a forestry worker watched a tall, dark-furred sasquatch with long hair on its neck and shoulders, walking off with a graceful stride, "as if it was very fit." (NAGATS: 10, 210) (map9)

(S86A) 1986, June: hikers spot a nine- or ten-foot-tall black creature and find a possible resting place of the creature, location not specified. (www.bfro.net)

(T88A) 1988, summer: wildlife biologist John Bindernagel and wife Joan, hiking with school group, finds fifteen-inch track at Lake Helen McKenzie, Strathcona Provincial Park, Vancouver Island. (NAGATS: 212) (track illustrated in figure 27) (map11)

(B88A) 1988: bulldozer operator who was pushing a road into the Nass Valley, north of Terrace, B.C. reported that as he has stopped to do a repair on his equipment, a snag (dead tree) "came hurtling across the road, thrown, not dropped." (TTR, no.91, p.13) (map6)

(S92A) 1992: Dennis Richards, Port McNeil, reported a sasquatch sighted walking up a beach. The location was Crease Island, eight miles northeast of Telegraph Cove. (NAGATS: 209.) (map9)

(VB92)   1992, summer: Mary Strussi, Courtenay, filed a report with John Bindernagel of nocturnal branch-breaking, sounds of an animal running back and forth and vocalizations including a "mournful bellowing." The events lasted all night during a camping trip near Comox Lake, central Vancouver Island. (NAGATS: 122, 227) (map11)
(B92A)   1992 or 1993: a Qualicum woman reported to John Bindernagel that she and another woman had been sleeping in a truck camper at a layby beside Sproat Lake, central Vancouver Island, during night front end of truck was lifted at least one foot and then dropped. When the women rushed outside to find cause, nothing seen. (NAGATS: 114, 224) (map11)
(N92A)   1992: Agate Beach, near Massett, Kelly Needham, second-hand report from a Massett woman of an unusual ten-foot diameter nest in forest, "not like a fallen eagle's nest." (map7)
(ST93)   1993, fall: Prince Rupert newspaper carried an article about two stranded fishermen who spent several days stranded on Banks Island, seventy miles south of Kitimat. Men saw a sasquatch walking on beach in evening, later found tracks where one had investigated their camp overnight. (submitted to BFRO.net June 27, 1998, © bfro.net) (map8)
(S94A)   1994, early June: in July, British tourist R. S. spoke with a resident beach vacationer at China Beach, southwest coast of Vancouver Island, who reported seeing a dark sasquatch running into forest from the beach. (© bfro.net) (map11)
(S94B)   1994, September: Tom Sewid described one of two sasquatches sighted on the shore of an island near Alert Bay, using the spotlight of his seine boat, the *Skidegate*. The chest was described as "like one and a half 45-gallon barrels." Dean Bolio witnessed. JoJo Christianson described strong smell. Sewid later heard sound of snags crashing after creature entered forest. (NAGATS: 114, 210, 215) (map9)
(S95A)   1995, early August: Jennifer Little reported (to Adrian Dorst) that she and a friend watched from a Zodiac inflatable off Vargas Island, near Tofino, Vancouver Island, as a sasquatch walked across a beach with typical sasquatch gait. (NAGATS: 41, 211) (map11)
(S95B)   1995, fall: dark brown sasquatch reported seen hiding behind a large boulder by forestry technician on mid-Princess Royal Island. (© bfro.net) (map8)
(SB96)   1996, spring: a second-hand report from Otto Winnig, Courtenay, related a logging mechanic at remote location near Harris Creek,

southwestern Vancouver Island experiencing two small stones thrown on truck hood as he worked on logging truck at night. After second stone, mechanic walked round vehicle to see "big, hairy, apelike man" stride quickly up bank. (NAGATS: 116, 225) (map11)

(V00A) 2000, spring: forestry professional Regan Dickins reported a second-hand account from a biologist friend who was at high elevation looking at rocks on Princess Royal Island in spring. Biologist reported hearing "some quite strange whooping and screaming going on in the distance." (Regan Dickins, pers. comm., 6/03/00) (map8)

(S02A) Read Island, 2002, summer: biologist John Bindenagel investigated report of a sasquatch sighted on the island.

(S02B) Port Alberni, Vancouver Island, 2002, summer: newspaper reported a seven-to eight-foot sasquatch with orange eyes sighted twice near Tofino at Long Beach and Radar Hill by Patrick and Arnold Frank. Weird howling reported around Indian Bay.

Reports in Appendix 3 are suggestive of a continuous population of unidentified hominids ranging from the B.C./Alaska border to Puget Sound. What is immediately obvious is the reported occurrence of sasquatches on Vancouver Island, Haida Gwaii (the Queen Charlotte Islands) and the myriad of islands up and down the coast. This suggests that geography poses few barriers to a primate such as the sasquatch, and that one could expect to find them on a remote Alaskan island just as easily as crossing a Vancouver Island road. In short, sasquatches have been observed traveling just about everywhere along the continent's northern coast and, if current sighting and track reports are any indication, they will find no difficulty in continuing to do so.

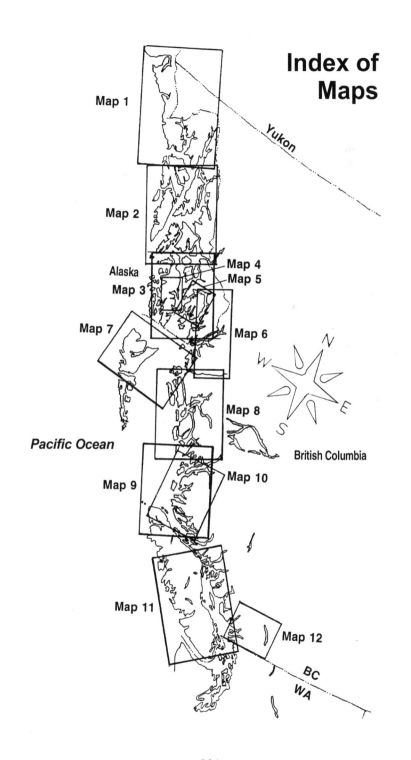

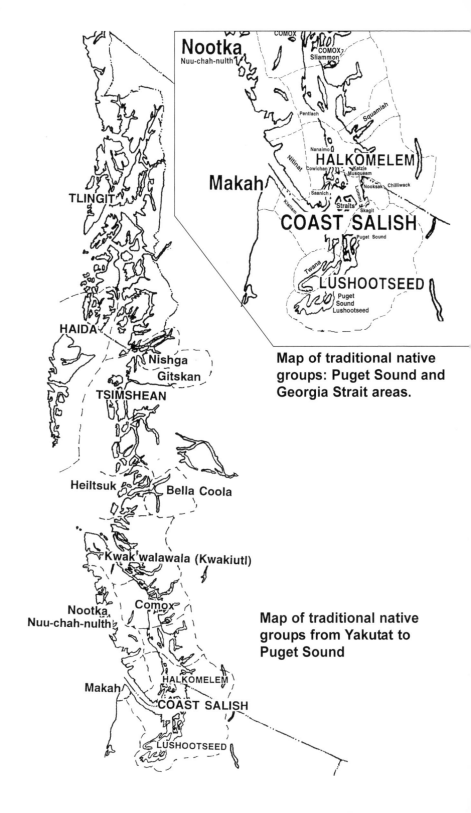

Map of traditional native groups: Puget Sound and Georgia Strait areas.

Map of traditional native groups from Yakutat to Puget Sound

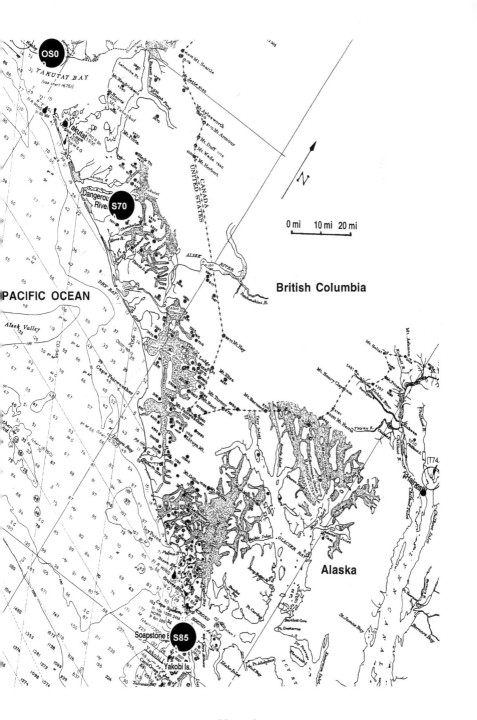

Map 1

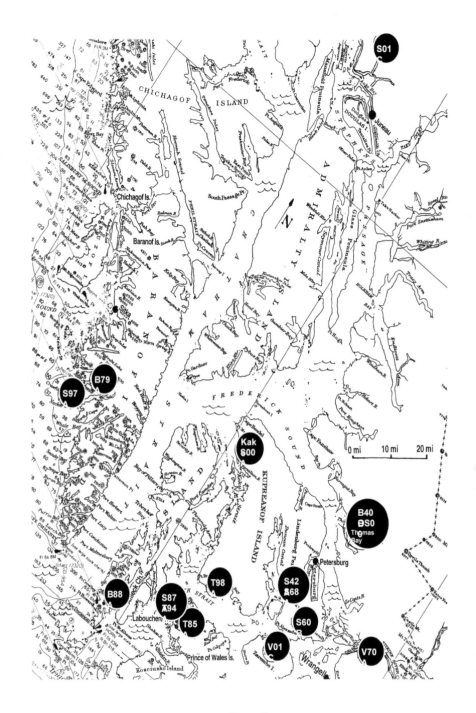

Map 2

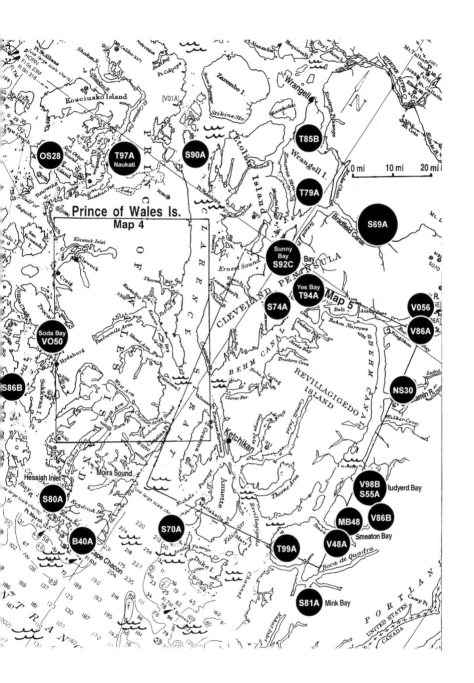

Map 3

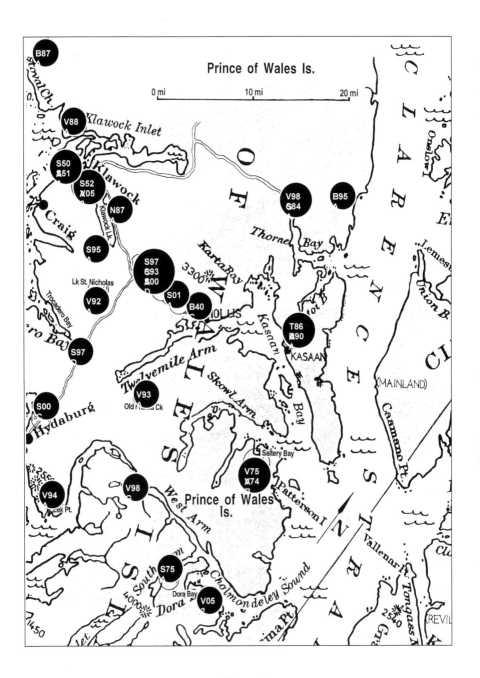

Map 4

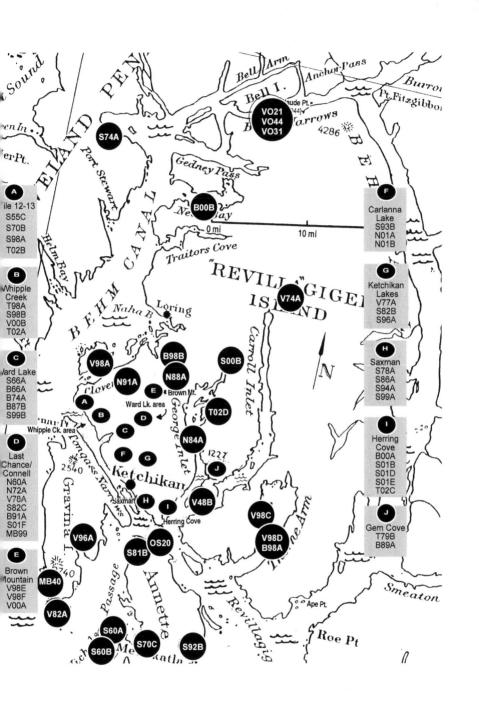

Map 5

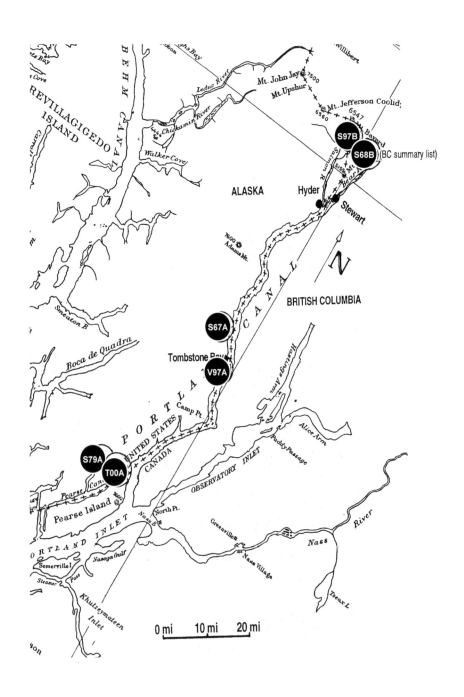

Map 6

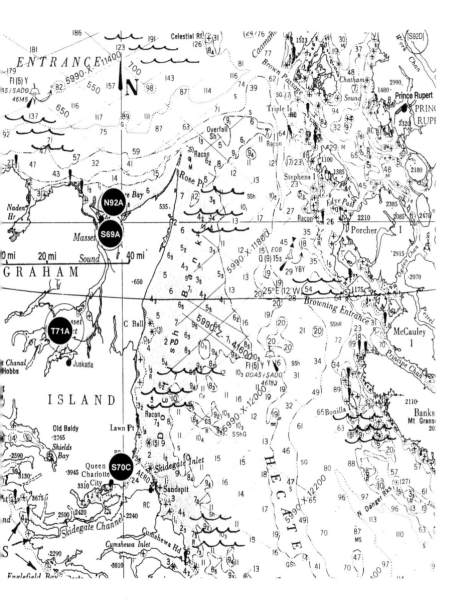

Map 7

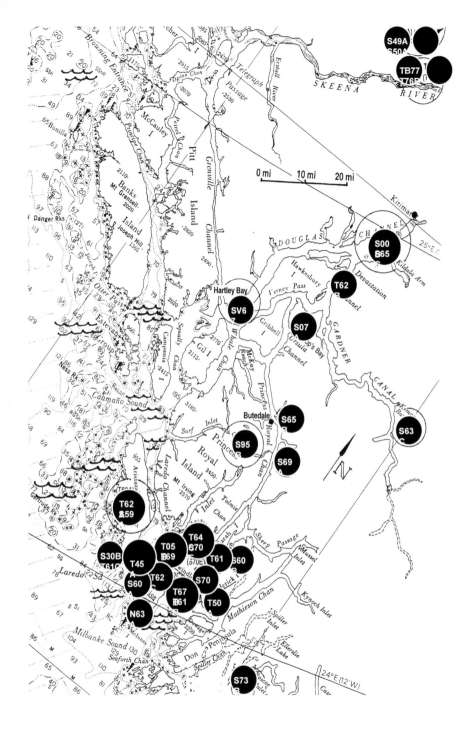

Map 8

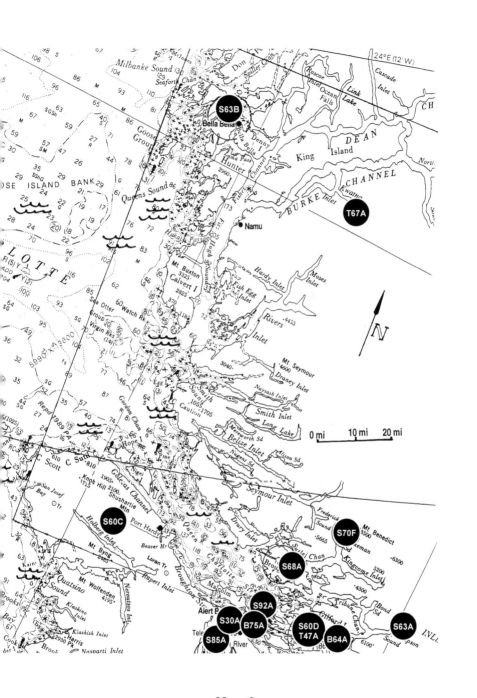

Map 9

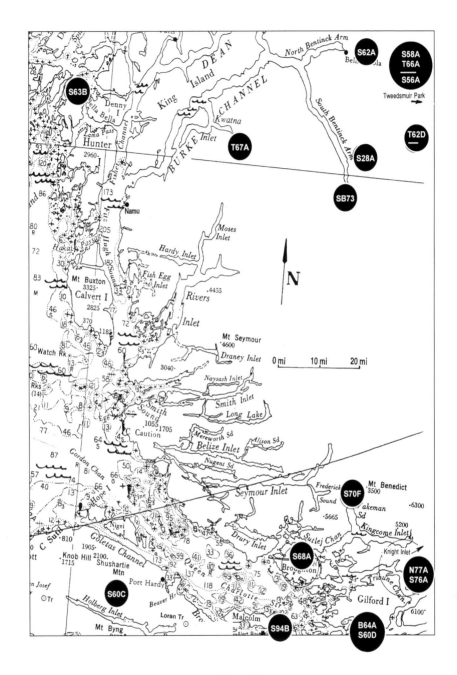

Map 10

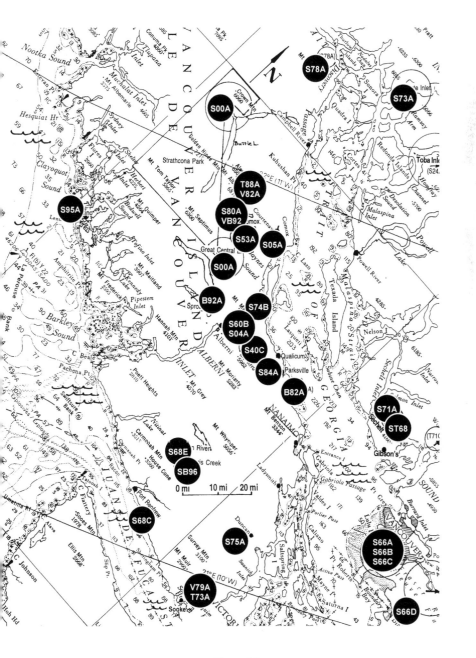

Map 11

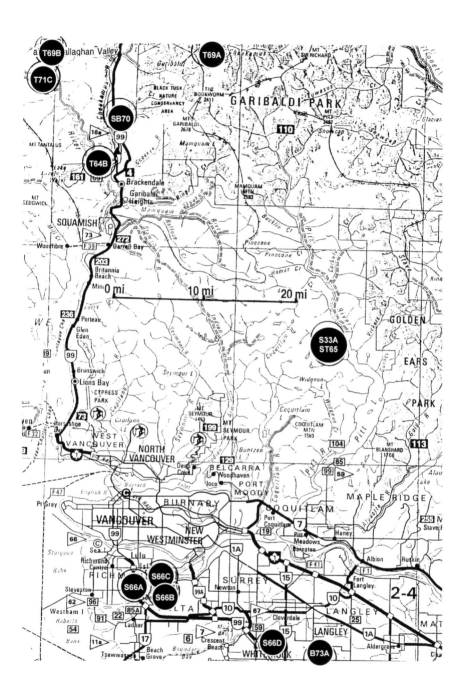

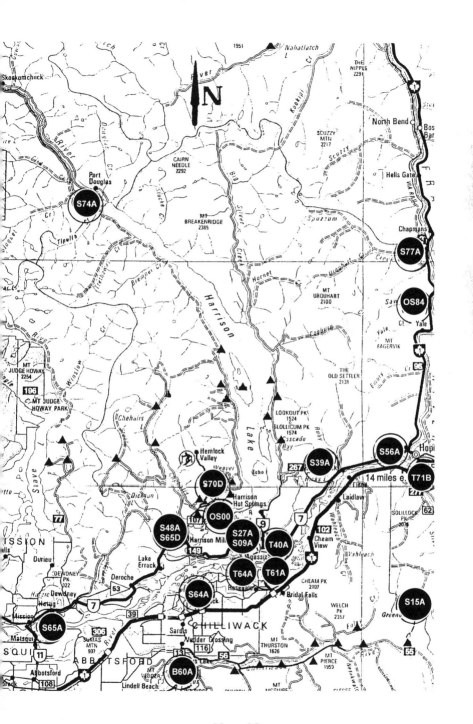

Map 12

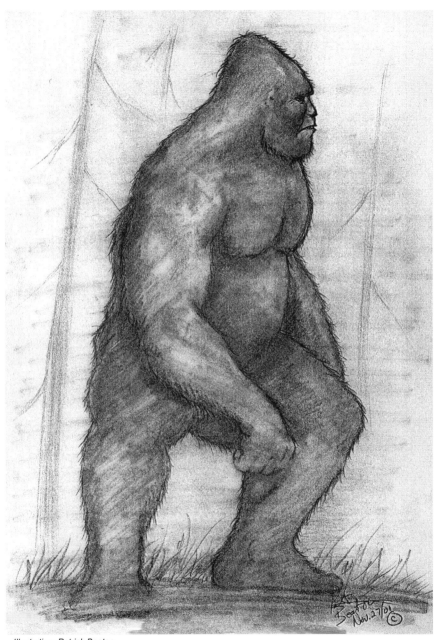

Illustration: Patrick Beaton

Bibliography

Barbeau, Marius, 1929, *Totem Poles of British Columnia*. Ottawa: National Museum of Canada Bulletin 61.

Bass, W., 1971, *Human Osteology: A Laboratory and Field Manual of the Human Skeleton.* Columbia: Missouri Archaeological Society.

Bayonov, Dmitri, 1997, *America's Bigfoot: Fact, Not Fiction.* Moscow, Russia. Crypto Logos Publishers.

Bindernagel, John, 1998, *North America's Great Ape: The Sasquatch.* Courtenay, B.C.: Beachcomber Books.

Boas, Franz, 1932, *Bella Bella Tales.* New York: The American Folklore Society.

Bord, Colin and Janet, 1982, *The Bigfoot Casebook.* Harrisburg, PA: Stackpole Books.

Brown, Albert, 1990, *As I remember-recollections from the house of raven.* Colleen and Company, Ketchikan, Alaska.

Byrne, Peter, 1975, *The Search for Bigfoot: Monster, Myth or Man.* Washington, DC: Acropolis Books.

—— 1976, *The Search for Bigfoot.* NY: Pocket Book.

Campbell, Bernard G., 1985, *Human Evolution.* 3d ed. New York: Aldine.

Ciochon, Russell, John Olson and Jamie James, 1990, *Other Origins: The Search for the Giant Ape in Human Prehistory.* New York: Bantam Books.

Ciochon, Russell and John Fleagle, editors, 1987, *Primate Evolution and Human Origins.* New York: Aldine.

Colp, Harry, 1997, *The Strangest Story Ever Told.* Petersburg, Alaska: Pilot Publishing. (originally published 1953, New York: Exposition Press.)

Dahinden, Rene and Hunter, Don, 1973, *Sasquatch.* Toronto: McClelland and Stewart.

Dixson, A. F., 1981, *The Natural History of the Gorilla.* New York: Columbia University Press.

Driver, Harold, 1967, *Indians of North America.* 2nd ed. Chicago: University of Chicago Press.

Duff, Wilson, 1952, *The Upper Stalo Indians of the Fraser Valley, British Columbia,* Anthropology in British Columbia Memoir No.1, Victoria, B.C.: British Columbia Provincial Museum.

Dunn, John A., 1995, *Sm'algax: a reference dictionary and grammar for the Coast Tsimshean language.* Juneau, AK: Sealaska Heritage Foundation.

Elmendorf, W..W., 1960, *The Structure of Twana Culture.* Pullman: Wahington State University Research Studies. Monograph Supplement 2.

Ferrell, Ed, 1996, *Strange Stories of Alaska and the Yukon.* Fairbanks, Seattle: Epicenter Press.

Fossey, Dian, 1983, *Gorillas in the Mist.* Boston: Houghton Mifflin.

Goodall, J., 1986, *The Chimpanzees of Gombe: Patterns of Behavior.* Cambridge: Harvard University Press.

Garza, Dolly, 1998, *Surviving on the foods and water from Alaska's southern shores.* Fairbanks, AK: University of Alaska, School of Fisheries and Ocean Sciences.

Green, John, 1968, *On the Track of the Sasquatch.* Agassiz, B.C.: Cheam Publishing (Reprinted by Ballantine Books, New York, in 1973.)

—— 1970, *The Year of the Sasquatch.* Agassiz, B.C.: Cheam Publishing Company.

—— 1973, *The Sasquatch File.* Agassiz, B.C.: Cheam Publishing.

—— 1978, *Sasquatch: The Apes Among Us.* Saanichton, B.C., Cheam Publishing Company and Hancock House Publishers.

Halfpenny, James, 1986, *A Field Guide to Mammal Tracking in North America.* Boulder, CO: Johnson Publishing.

Halpin, Marjorie and Ames, Michael, editors, 1980, *Manlike Monsters on Trial: Early Records and Modern Evidence.* Vancouver, B.C.: University of British Columbia Press.

Heller, Christine, 1989, *Wild Edible and Poisonous Plants of Alaska.* Fairbanks, AK: University of Alaska Cooperative Extension Service.

Heuvelmans, Bernard, 1959, *On the track of Unknown Animals.* NY: Hill and Wang.

Hewkin, James A., 1990, "Sasquatch investigations in the Pacific Northwest", *Cryptozoology*, Vol.10: 76–78

Jackson, W..H., 1974, *Handloggers.* Anchorage, AK: Alaska Northwest Publishing Company.

Krantz, Grover, 1971, "Sasquatch handprints." *Northwest Anthropological Research Notes* 5(2):145–151. (Reprinted in *The Scientist Looks at the Sasquatch.*)

—— 1972a, "Anatomy of the sasquatch foot." *Northwest Anthropololical Research Notes* 6(1):91–104. (Reprinted in The Scientist Looks at the Sasquatch.)

—— 1972b, "Additional notes on sasquatch foot anatomy." *Northwest Anthropological Research Notes* 6(2):230–231.

—— 1981, *The Process of Human Evolution.* Cambridge, MA: Schenckman.

—— 1983, "Anatomy and dermatoglyphics of three sasquatch footprints." *Cryptozoology* 3:131–134.

—— 1992, *Big Footprints: A Scientific Inquiry into the Reality of Sasquatch.* Boulder, CO: Johnson.

—— 1987, "A reconstruction of the skull of Gigantopithecus and its comparison with a living form." *Cryptozoology* 6:24–39.

Kricher, John C., and Morrison, Gordon, 1993, *Peterson Field Guides-Ecology of Western Forests.* Boston, MA: Houghton Mifflin.

Levi-Strauss, Claude, 1988, *The Way of the Masks,* (trans.: Sylvia Modelski) Vancouver, B.C.: Douglas and McIntyre.

Lockley, Martin, 1999, *The Eternal Trail: A Tracker Looks at Evolution.* Cambridge, Mass: Perseus Books.

Mack, Clayton, 1993, *Grizzlies and White Guys.* Madeira Park, BC, Harbour Publishing.

MacNair, Peter, Robert Joseph and Bruce Grenville, 1998, *Down from the Shimmering Sky.* Vancouver: Vancouver Art Gallery and Douglas and McIntyre.

Markotic, Vladomir (ed) and Krantz, Grover (associate ed.), 1984, *The Sasquatch and Other Unknown Hominoids.* Calgary, Alberta: Western Publishers.

Maud, Ralph, 1982, *A Guide to B.C. Indian Myth and Legend.* Vancouver, BC: Talonbooks.

McIlwraith, T..F. 1948, *The Bella Coola Indians.* Volumes I and II. Toronto: University of Toronto Press.

Mozina, Jose, 1972, *Noticias de Nutka.* ed. Iris H. Wilson. Seattle: University of Washington Press.

Napier, John, 1972, *Bigfoot.* NY: Berkeley.

Napier, John and P. H. Napier, 1985. *The Natural History of the Primates.* Cambridge, Mass: M.I.T. Press.

O'Clair, Rita, Robert Armstrong and Richard Carstensen, 1992. *The Nature of Southeast Alaska.* Anchorage: Alaska Northwest Books.

Park, Michael Alan, 1996, *Biological Anthropology.* Mountain View, CA, London, Toronto: Mayfield Publishing.

Perez, Danny, 1988, *Big Footnotes: A Comprehensive Bibliography Concerning Bigfoot, The Abominable Snowmen and Related Beings.* Norwalk, CA: D. Perez Publishing. Inc.

Peterson, Roger Tory, 1990, *A Field Guide to Western Birds.* Boston, MA: Houghton Mifflin.

Pojar, Jim and MacKinnon, Andy, 1994, *Plants of the Pacific Northwest Coast.* Vancouver, B.C.: Lone Pine.

Rigsby, Bruce, 1977, "Some Pacific Northwest Native Language Names for the Sasquatch Phenomenon." in *The Scientist Looks at the Sasquatch (I)*, edited by R. Sprague and G.Krantz, pp31-38. Moscow, ID: University of Idaho Press.

Robbins,C. S., Bruun, B. and Zim, H. S. *A Field Guide to the Identification of Birds of North America.* 1983, NY: Golden.

Rohner, R. P. and Rohner, E. C., 1970, *The Kwakiutl Indians of British Columbia.* New York: Holt, Rinehart and Winston.

Place, Marian T.,1974, *On the Track of Bigfoot.* NY: Dodd Mead.

Relethford, John H., 1994, *The Human Species: An Introduction to Biological Anthropology.* Mountain View, CA: Mayfield Publishing.

Sanderson, Ivan, 1961, A*bominable Snowmen: Legend Come to Life.* NY: Chilton.

Sapir, E. A. and M. Swadesh, 1939, *Nootka Texts: Tales and Ethnographic Narratives.* Philadelphia: Linguistic Society of America, University of Pennsylvania.

Schaller, George, 1963, *The Mountain Gorilla: Ecology and Behaviour.* Chicago: University of Chicago Press.

———1964, *The Year of the Gorilla.* Chicago: University of Chicago Press.

Shackley, Myra, 1983, *Still Living?: Yeti, Sasquatch and the Neanderthal Enigma.* NewYork: Thames and Hudson.

Sheldon, Ian and Hartson, Tamara, 1999, *Animal Tracks of Alaska*. Edmonton, AB, Seattle, WA: Lone Pine Publishing.

Simons, E. L. and Pilbeam, D., 1971, "A gorilla-sized ape from the Miocine of India." *Science* 173 (1991): 23–27)

Slate, B. Ann and Alan Berry, 1976, *Bigfoot*. New York: Bantam Books.

Spradley, James P., 1979, *The Ethnographic Interview*. New York, Holt Rinehart and Winston.

Sprague, Roderick, and Grover, Krantz, editors, 1977, *The Scientist looks at the Sasquatch (I)*. Moscow, ID: University of Idaho Press.

—— 1979, *The Scientist Looks at the Sasquatch (II)*. Moscow, ID, University of Idaho Press.

Steenburg, Thomas, 1990, *The Sasquatch in Alberta*. Hancock House Publishers: Surrey, BC.

—— 1993, Sasquatch/Bigfoot, *The Continuing Mystery*. Hancock House Publishers: Surrey, BC.

—— 1996, *In Search of Giants*. Hancock House Publishers: Surrey, BC.

Strassenburgh, Gordon, 1984, "The crested *Australopithecus robustus* and the Patterson-Gimlin film." *The Sasquatch and Other Unknown Hominoids*, edited by V. Markotic and G. Krantz, pp 236–238. Calgary, Alberta: Western Publishers.

Suttles, Wayne, 1987, *Coast Salish Essays*. Vancouver, BC: Talonbooks.

Swanton, John, 1909, *Tlingit Myths and Texts*. Bureau of American Ethnology Bulletin 39.

Weidenreich, Franz, 1945, "Giant early man from Java and south China." *Anthropological Papers of the American Museum of Natural History* XL, No. 1.

Whinney, Stephen, 1989, *The Audubon Society Nature Guides–Western Forests*. NY: Alfred A. Knopf.

Index

GENERAL INDEX
(see also Index of Places, Index of Names)

Acromegaly, 264
Aggression. *see* Behaviors
Apes compared to sasquatches: displays, 215, 273; generally, 14, 126–27, 182, 212, 253, 271–76, 286; nest building, 117, 125, 237, 244, 246–47
Athabascan First Nation, 126
Atlakwis, 149
Australopithecus robustus, 53, 271–72, 276–77

"Bay of Death," 118
Bears and sasquatches, 91–93, 198, 209–10, 232–35, 246–47
Behaviors: abducting humans, 107–15, 118, 125–26, 150–51, 154, 156–57, 163; about, 80–81, 95–97, 255, 273–74, 281, 286; boarding boats, 51, 54, 231; breaking branches/logs, 44, 65, 120, 147, 206, 240, 254, 273; circling, 121–22, 156, 195, 197; entering camp shelter, 40; harassing humans, 46, 65, 218–20; jumping into cover, 19, 36, 69, 78; nocturnal, 30, 56, 151, 163–64, 257; play, 40, 114, 146–47, 220, 224–26, 255, 277; running alongside a vehicle, 67–68, 220; shielding eyes with arms, 30, 52; stacking bones neatly, 228; stealth, 172, 225, 232–34; territorial, 81, 95, 220, 248–52, 255, 273; theft/stealing, 40, 159, 163, 178, 227–28, 231; threat displays, 212, 215, 248; trees and, 99, 119, 156, 248–52; vandalism, 65, 69–70, 188, 203, 206, 217, 221–23; watching, 34–35, 45–46, 52, 58, 60, 76, 84, 87, 92, 96–97, 99–100. *see also* Rocks; Tool-using; Vocalizations
Behaviors, feeding: about, 79–81; chasing game, 90, 226; digging, 20–21, 85; drinking with cupped hand, 45; foraging, 79, 148, 151, 189–90, 229, 237; hanging deer kills, 229; pulling bark off trees, 43, 78, 207, 248; theft of game, 159, 178, 227–28, 231; washing and stacking roots, 229. *see also* Food(s)
Behaviors, noise-making: about, 210, 215; crashing through vegetation, 65, 206, 255; rapping, wood-on-wood, 34, 147, 201–2, 205–6, 214, 242; rattling door knobs, metal cans, 65; rocks, clacking together, 206, 223; "splooshing" sounds, 254
Bella Bella First Nation, 150–51
Bella Coola First Nation, 100–101, 147–49
Bergman's Rule, 124
Bigfoot Bulletin, 34
Bigfoot Digest, 62
Bigfoot Researchers Organization, 32, 290
Bipedalism. *see* Body position; Gait
Body, skin on, 100, 268–69
Body condition, 20–21, 73, 119
Body position: bending over, 20–21, 79; crouching, 30, 36, 45, 76; forward slump, 73; kneeling, 99; sitting down, 113; sleeping or resting, 48–49; squatting, 72, 76. *see also* Gait
Body shape: about, 94; apelike, 52; barrel chest, hump on back, 113; broad hips and buttocks, 73; chest, 50-55", waist 36-38," 113; humanlike, 20–21, 39, 58, 84, 86–87; legs, 33, 42, 79, 90, 264; man-shaped, 21, 30, 42, 51, 67, 72,

99, 102; man-sized, 22–23; neck, 39–40, 47–50, 62, 72, 113; shoulders, 22–24, 39–40, 48–49, 52, 63, 113; tall with broad shoulders, 23–24. *see also* Feet; Hands; Head shape; Height estimates
Bone material, 263–70, 284

Cameras, infrared, 288
Caves. *see* Nests and dens
Chasing, 90, 119, 226
Children (sasquatch): about, 32; male and female, 109–14; play of, 111, 114; toddler sighting, 82–84; tracks of, 167
Chimpanzees, 200, 215, 257, 271–72, 274
Claw marks, 122, 240
Coast Salish First Nation, 155–60
Color: about, 94–95, 124–25, 219; black, 28, 39, 48, 51–52, 66, 76, 80–82, 84, 86–87, 90–91, 151, 155, 175–76, 225, 268; black, grayish brown, 171; bluish gray, 102; brown, 29, 33–34, 234; brown, dark, 22, 28–29, 40, 45, 48, 51, 54, 78, 81, 84, 187; brown, light, 18–19, 22–23, 82, 99, 170, 261; chestnut, reddish brown, 79; dark, 25, 27, 30–33, 35–36, 43, 45, 67, 84, 158, 220; grayish, 58; reddish-brown, 60, 66; tan, dark, 63. *see also* Hair (or fur) on body

Comox First Nation, 154
Conclusions: on abductions, 157; cultural comparison of traits, 160–66; on curiosity, 35–36, 41, 46, 56, 95; on existence of sasquatch, 11–12, 14, 291–92; on foods, 77–78; on hibernation, 88; on Ostman's experience, 117–18; on photographing sasquatches, 57, 288–89; on range of characteristics, 126; on sightings, 94–97, 291–92; on smaller sasquatches, 32; on stalking, 46; on swimming, 64; on tearing bark off limbs, 79; on thumbs, 55
Condition. *see* Body condition
Cross-cultural comparison of traits, 160–66
Crouching, 30, 36, 45, 76
Curiosity: about, 35–36, 41, 46, 56, 95; of sasquatches, 40, 62, 220, 225–26

Demeanor: about, 96; beligerent, 95; friendly, 100; gentle, 47–48, 54–57, 125, 212–13, 224; not threatening, 41, 52; shy, reclusive, 15, 95, 153, 226, 285–86. *see also* Curiosity
Dens. *see* Nests and dens
Dermatoglyphics, 182
Devils, black, hairy, small, 119, 123
Displays, intimidation. *see* Behaviors
Disposition. *see* Demeanor

Distance from humans: 3-6 yards, 20–21, 23–24, 48–50, 52, 60; 10-17 yards, 21–23, 39, 43, 87, 91; 20-30 yards, 36, 42, 176; 40 yards, 29, 45, 51, 63; 50-75 yards, 18–19, 24–25, 29, 86, 88, 99–100; 100-150 yards, 28, 63, 71, 76; 200-500 yards, 34, 37, 82, 85, 92
DNA analysis, 259, 275–76, 289. *see also* Research
Dogs, 44, 121, 126, 283
Drinking with cupped hand, 45
Dwarves of Pybus Bay, 124

Eagle nests, 246–47
Eyes: almond-shape, 60; light brown with large, dark pupils, 72; reddish reflection, 30; reddish-whitish reflection, 52, 225; round, 43, 58, 62–63; whitish reflection, 43, 88; wide open, deep-set, 151

Facial features. *see also* Facial skin
Facial features: almost human, 225; apes compared to, 277; brow ridge, 48, 74, 151; cheekbones, 60; ears, 100; face shape, 40, 52, 62, 72, 100, 277; flat, 63; hair, 29, 48–49, 63, 72, 113; jaws, 48, 78, 113, 265–67; mouth, 62, 72, 74, 100, 113; nose, 48, 60, 72, 100, 150; teeth, 155, 255, 270, 277, 284. *see also* Eyes

353

Facial skin: ash gray colored, 52; dark brown or black, wrinkled, with patches of light, 72; leathery, dark gray, 48; muddy, tan-green, 62; tan, wrinkled, 63
Families of sasquatches, 109–14
Fecal samples, 262
Feet: about, 75–76, 167–70, 287–88; dermal ridge pattern, 182–83, 280; flexible, bending inward, 72; heels sticking out, 73; matted fur on, 187–88; partial webbing, 51, 113; remains found, 218–19, 267–70; soles, 29, 113; toenails, large curved, 155. *see also* Toes; Tracks
Females: breasts, pregnant, 72–73; *dsonoqua*, 151–53; longer hair on head, wide hips, 113; meek, 72–73; *qelqelitl*, 157
First Nations, 161–66. *see also specific First Nations*
Food(s): about, 77–78; bear, 232–33; berries, 46; ducks, 234, 274; grass with sweet roots, 112–13; grubs, 43–44, 78, 100, 207, 248; macaroni, 110; mice & berries, 235; nuts, 111; prunes, 110; salmon, 26; snuff, 114; vegetarianism, 125; venison, 226, 233–34. *see also* Behaviors, feeding
Foraging. *see* Behaviors, feeding
Fur. *see* Facial features: hair; Hair on head; Hair (or fur) on body

Gait: lumbering, 35, 40; moving very fast, 27–29; running very fast, 18–19, 24–25, 67, 91–92, 220, 226; walking, bouncing, 30; walking, manlike, 21, 30, 37, 43, 45, 70, 84–85, 99, 102; walking, swaying, 43, 73, 113; walking, swinging arms, 22–25, 29, 36, 47, 67, 81, 88; walking, toddlerlike, 82–83; walking fast, 26, 81, 176, 186
Gentleness. *see* Demeanor
Gigantopithecus, 13–14, 168, 271–72, 275–80, 291. *see also* Apes
Gorillas. *see* Apes

Haida First Nation, 17–18, 66–67, 81, 140–44, 248
Hair on head: long, 49, 60, 113; long, like a mane or cape, 31–32, 37, 72; long and matted, 20, 39–40
Hair (or fur) on body: about, 164, 278; coarse, 66, 119; hemp-like, 62; long, 20, 48, 72, 76–77, 81, 92, 261; matted and foul, 187; shiny, 25; short and thick in places, 113; between toes, 268–69. *see also* Color; Facial features: hair; Hair samples
Hair samples, 241, 243, 245, 259–62
Hands: communication with, 112; fingernails, 113, 155; hair on, 78, 151; large, 54, 113; palms, 29, 54, 78, 113; thumbs, 54–55
Head shape: larger than human, 60; manlike, 22–23, 155; pointed, 60, 63, 72; slope in front, rounded in back, 74, 113; small compared to body, 100. *see also* Body shape; Hair on head
Height estimates: 3-4 feet, 82–83; 5 feet, 26, 36; 6-7 feet, 18–19, 24–25, 52, 80, 85–86, 211; 7 feet, 19, 22–24, 29–30, 33, 39, 43, 46–47, 66–67, 78, 112–13, 158–59, 175–76, 187, 220; 7-8 feet, 28–29, 42, 51, 73, 76, 211; 8 feet, 30, 81, 84, 99, 113, 158; 9-10 feet, 28, 45, 89–92, 226; about height, 160–61, 163–64, 280–81; man-sized, 22–23, 25, 82
Heiltsuk First Nation, 149–51
Hibernation, 88
Homo erectus, 276–80
Homo neanderthalensis, 271–72

Indians. *see* First Nations
Invisibility, 159–60

Juveniles. *see* Children (sasquatch)

Kwakiutl First Nation, 151–53

"Land otter people," 137–39
Language, 212

Lummi First Nation, 158–59, 235

Map coding guide, 18
Masks and wood carvings, 138–39, 146–47, 150–53, 164, 277
Moose and elk, 253
Musqueam First Nation, 157

Names, local: *ba'oosh,* 9, 145–47; *ba'wes/ba'wis,* 9, 145; *boqs,* 147–49; *buk'wus,* 145, 151–55; *c'amek'wes,* 158–59; *c'iatqo,* 159–60; *dzonoq!wa* or *dsonoqua,* 151–52; *gagiit,* 9, 140–44; hairy man, 18, 28–29, 143; *K!a'waq!a,* 150–51; *kushtakaa,* 9, 17, 137–40; *kwai-a-tlat,* 155–57; "land otter people," 137–39; *mai-a-tlatl,* 154–55; *matlox,* 154–55; Oh-mahs, 10; *papay'oos,* 155–57; *pkw's,* 150–51; sea man, 61–62; *sesq'ec* and *qelqelitl,* 157; *sesxech,* 9; *squee'noos,* 155–57; *suhsq'uhtch,* 157–58; wild man, 151–53
Nanaimo First Nation, 126, 155–57
Naxalk First Nation, 100–101, 147–49
Neanderthal, 271–72
Nests and dens: about, 117, 218, 237–38, 279; branches and boughs, 238–40, 244–46, 261; canopies, 237–38; cave, 48–50; glacier, 102; mine shaft, 50, 239; shelf on mountainside, 110; shredded bark,
logs, branches, 245–46; woven materials for, 110, 125, 240, 243–44
Nisga First Nation, 145
Nootka First Nation, 154–55
Nuu-chah-nulth First Nation, 154–55
Nuxalk First Nation, 147–49

Odor: animal-smell, 199, 238; musk and earth, 40; of open sores and bodies, 119; rank, 20–21, 25, 67, 195, 225; wet hair or fur, 52, 78, 225
Orangutans. *see* Apes

Paranthropus, 271–72, 278–79
Photographing sasquatches, 57, 288–89
Piloerection, 213
Pokmis, 155
Posture. *see* Body position

Quadrapedal foraging, 79

Ramapithecinae, 275–76. *see also* Apes; *Gigantopithecus*
Research, 10–11, 75, 93–94, 182, 235, 258, 267, 284–85. *see also* DNA analysis; Hair samples
Rocks: clacking together, 206, 223; stacking/building cairns, 80, 255–58; as thermal traps, 257; throwing, 44, 145, 222–24
Running. *see* Gait

Sagittal crest. *see* Facial
features
Sasquatches: name origin, 9; as unknown species, 280–81
Shape. *see* Body shape; Facial features: shape; Head shape
Shootings by humans, 46–47, 148–49
Sightings safety, 290
Skeleton, 264
Smell. *see* Odor
Sounds, non-vocal. *see* Behaviors, noise-making
Sounds, vocal. *see* Vocalizations
Southeast Sasquatch Research Group of Ketchikan, 79
Speed. *see* Gait
Stealing. *see* Behaviors: theft/stealing
Straddle, 178; 186
Strength, 217–20, 229–30
Stride, 29, 168–69, 180, 186–88. *see also* Gait
Swimming, 51, 58, 60, 62–64, 66, 163
Synopses. *see* Conclusions

Throwing: 4-5' log, 216; 600-pound furnace, 26, 230; about, 231; pine cones, 146–47, 223; rocks, 44, 145, 222–24; snow balls, 216–17
Tlingit First Nation, 17–18, 118, 124, 137–40, 224
Toes: dermal prints of, 288; hair between, 268–69; partial webbing, 51, 113; spread and arch of, 176–77; variable toe imprinting,

355

183. *see also* Feet
Tool-using: about, 277–78; digging for claims with stick, 85; rocks as thermal traps, 257; route-marking, 248–52, 255, 257; sticks for springing traps, 66, 217; weaving, 110, 117, 125, 240, 243–45. *see also* Behaviors
Tracks: 13-16" long, 74, 80, 174; 15" long, 40, 80, 167, 176, 179–80, 188, 208; 16" long, 77, 167, 173, 181, 227; 18" long, 170, 176–77, 183–84, 261; 19" long, 91; 20" long, 158, 182; about, 189; claw marks, 122; in grass, 44; humanlike, 91, 181, 227; large, 27, 36, 145, 199; mingled with dog's, 121; in mud, 167, 176; in sand, 179–80, 182; in snow, 27, 36, 170–72, 174–77, 181, 183, 186–89, 199; in soft brown soil, 173; in wet silt, 189. *see also* Feet
Traits, cross-cultural comparison, 161–66
Transformation, 138–42, 160, 163
Tribes. *see* First Nations
Tsimshean First Nation, 145–47
Twana First Nation, 159–60

Upper Stalo First Nation, 157–58

Vandalism. *see* Behaviors
Vocalizations: about, 95, 117–18, 191–92, 203, 209–12, 279–80, 286; attempt to communicate, 197; baby crying, 137, 142, 254; barking, 196; breathing, 40, 107, 119, 198; calls, loud/prolonged, 195, 200, 202, 207; calls, multiple, 192–94; chattering, 109–14, 125, 196–97, 208, 211; chuckling, 142, 199, 208; cries, 119, 204; grunting, 73, 147, 195, 211; "Haii Haii Haii," 101; hooting, 65–66; howling, 155, 208; "oook," 112; roaring, 147–48; screams, loud, 24, 88, 91, 119, 142, 181, 186–87, 196–99, 201–4, 208, 217–18; screams, loud, angry, 204–5, 207; screams, multiple, 201–2, 207, 209; speaking loud, 151; squeak/moan, 120; squeals, 114; wailing, 196; whinnying, 142; whistles, multiple, 202; whistling, 137, 151, 159, 164; whooping, 65–66, 200; yelling, 119, 191, 197, 203, 207 "The voice of Claude Point," 192–94

Walking. *see* Gait
Washington State Sasquatch Research Group, 288, 290
Western Bigfoot Society, 290
Wood carvings. *see* Masks and wood carvings

Zinjanthropus, 53

Index of Places

ALASKA
about the coast, 86–87, 98, 118
Admiralty Island, 124
Affleck Channel, 221–22
Agate Beach, 246
Annette Island, 18–19, 81–84, 145–47, 225–26, 263–64
Baranof Island, 45–46, 54–56
Bartholomew Creek, 265
Bear Mountain, 47–48
Bear Valley, 48–50
Betton Island, 204
Bishop's Island, 196–97
Blind Slough, 33
Boca de Quadra Inlet, 180
Bostwick Inlet, 202
Bradfield Canal, 34–36
Brown Mountain, 207
Bugge's Beach, 21
Burke Channel, 149
Cape Flattery, 62
Carlanna Lake, 175, 254–55
Carroll Inlet, 195–96, 198, 226–27
Chomondoley Sound, 79, 195
Clarence Strait, 61–62
Claude Point, 192–94
Connell Creek, 23–24
Connell Lake, 42–44, 199, 218, 267–70
Cottonwood Slough, 198
Cowboy Camp, 84
Crab Bay, 82
Dall Bay, 200
E. Behm Canal, 180
Eek Inlet, 202
Exchange Cove, 208

356

Flicker Creek, 174
Forrest Lake, 46–47
Gem Cove, 173, 227–28
George Inlet, 173, 182
Gertrude Lake, 207
Granite Basin, 199
Gravina Island, 84, 200, 202
Green Monster Mountain, 207
Haida lands, 140–44
Haines, 170, 261
Harriet Hunt Lake, 69, 217, 232–34, 244–45
Hecata Island, 123
Herring Cove, 24–26, 189, 230
Hessiah Inlet, 80–81
Hidden Glacier, 87–88
Hidden Inlet, 85–87
Hollis, 217
Hollis-Klawock Hwy., 27–28, 30–32
Hoonah, 170
Hydaburg, 29–30, 67–69, 207
Hydaburg, south of, 59–61
Hydaburg Hwy., 28–29
Hyder, 87–88, 197
Juneau, west of, 52–53
Kake, 36
Kasaan, 175, 228
Ketchikan: 10 mi. northwest of, 178–79; 10 miles south, 230; 14 mi northeast, 187–88; 20 miles NW, 205; 28 air mi. southeast, 179–80; east of, 76–78; near, 58–59
Klawock, 197, 201
Klawock-Hollis Hwy., 27–28, 30–32
Klawock Lake, 41–42, 65–67, 217, 239–44, 248–52, 256, 262
Klawock Mountain, 256

Kupreanof Island, 33–34, 179
Labouchere Bay, 26–27, 176–77, 179
Lake Creek, 196
Lake St. Nicholas, 202
Last Chance Campground, 23–24, 41–42, 199
Little Totem Bay, 179
Lynn Canal, 170
Mabel Island, 141
Mainland, south, 84–88
Malaspina Glacier, 101–2
McFarland Islands, 59–61
Metlakatla, 18–19, 82–84
Minerva Mountain, 47–48
Mink Bay, 87
Misty Fjords area, 87, 89–91, 194–95, 204, 234, 265–67
Muddy River, 121
Naukati, 177
Nichols Passage, 84
Ohmer Creek, 33
Old Frank's Watershed, 202
Panhandle, northern, 33–38
Patrick Lake, 174
Pearse Canal, 85–87
Pearse Island, 181
Petersburg, 33, 36, 118–23, 216
Point Baker, 174
Point Davis, 81
Point Sykes, 180
Portland Canal, 84–85, 181, 191, 203
Prince of Wales Island (*see specific cities or locations*)
Prince William Sound, 37–38

Pybus Bay, 124
Red Bay, 174
Reef Island, 181
Revilla' Island (*see specific cities or locations*)
Rudyerd Bay, 89–91, 204
Salmon Lake, 256
Salmon River, 197
Saltery Cove, 199
Saxman, 21–22, 45, 78–79
Second Waterfall Creek, 245
Signal Creek Campground, 22–24, 39–41
Sitka, 30-40 mi. south, 45–46, 54–56
Skowl Arm, 199
Slide Ridge, 238–39
Smeaton Bay, 194–95, 265–67
Soapstone Bay, 92–93
Soda Bay, 196
South Alaska Mainland, 84–88
Southeast Alaska, 137–40
Swan Lake, 198
Taiyasanka Inlet, 170, 261
Thomas Bay, 118–23, 126, 216
Thoms Lake area, 170–72
Thorne Arm, 222–24, 257–58
Thorne Bay, 67, 204–5, 220, 228–29
Toba Inlet, 279
Tombstone Bay, 84–85, 203
Tongass Hwy., 19–21, 91
Trocadero Bay, 28–29
Unuk River, 196, 200

Upper Ward Creek,
  23–24, 267
Upper Wolf Lake,
  232–34
W. Behm Canal, 181
Walker Channel, 45–46
Ward Lake, 22–24,
  39–44, 69–70, 206,
  218–20, 223–25,
  237–39
Whipple Creek, 91,
  178–79, 183–89, 207
White River, 181–82
Wild Man Cove, 195
Wrangell area, 33–36,
  71–75, 170, 174, 179,
  198
Yakobi Island, 92–93,
  234–35
Yakutat, 46–47, 101–2
Yes Bay, 181
Zarembo Island, 208–9

BRITISH COLUMBIA
  about the coast, 98–101
Bay of a Thousand
  Islands, 147–48
Bella Bella, 147–51
Bella Coola area,
  98–101, 147–49
Cultus Lake, 224
Dodd's Narrows, 156
Forbidden Plateau, 154
Fraser Valley, 157–58
Graham Island, 140–44
Jacobsen Bay, 99
King Island, 149
Kitimat, 145–47
Kwakiutl lands, 151–53
Kwakwalawala, 150–51
Kwatna, 148–49
Massett, 246–47
Namu, 147–49
Port John on Evans
  Arm, King Island, 149
Prince Rupert Terrace,
  145–47
Queen Charlotte
  Islands, 246–47
South Bentick Arm,
  99–101
Stewart, 87–88
Strathcona Park, 201
Toba Inlet, 102–18
Vancouver, 157–58
Vancouver Island,
  154–57, 201
Work Channel, 62–63

WASHINGTON
Hood Canal/Puget
  Sound, 159–60
Lummi area, 158–59
Nooksack, 235

Index of Names

Abbott, Thomas, 170,
  261
Allen, Henry, 159–60
Ame, Michael, 211
Armstrong, Robert H., 93
Babich, Lindsey, 62
Benson, Richmond, 248
Berry, Alan, 210–12
Bindernagel, John, Ph.D:
  about, 12; on aggressive
  behavior, 215, 218, 224,
  230–31; apelike whoop-
  ing heard by, 201; on
  belief in sasquatch, 11;
  on hair, 259; on nest
  building, 237; photo of
  track by, 169; on stack-
  ing rocks, 255
Boas, Franz, 151–52
Bob Titmus, 33
Bord, Colin and Janet,
  125
Bradshaw, Fred, 36, 46,
  257
Brown, Albert J., 144,
  248
Burns, J.W., 9, 124–25
Byrne, Peter, 59–61, 125

Calhoun, Mickey, 28
Cartensen, Richard, 93
Carter, Nick, 57
Chilcutt, James, 182,
  280, 288
Cogo, Eddie, 141
Cogo, Robert, 140–41
Colp, Harry, 118–23, 126
Crowe, Ray, 37, 59–61
Dahinden, Rene, 10, 102,
  235
Dickinson, Regan, 290
Dixson, A.F., 273
Dunn, John A., 145
Durgan, Robert, 175, 228
Durgan, "Tink," 85–86
Edenshaw, James, 48,
  277
Edenshaw, Stan, 28
Fahrenbach, W. Henner,
  37, 259–61, 292
Farrell, Ed, 102
Fisher, Thomas J., 91–93
Fossey, Dian, 14
Goodall, Jane, 271, 273
"Grandma Charlie," 125
Green, John: about,
  10–11; on aggressive
  behavior, 230–31;
  "Errol" story, 57–59; on
  face types, 277; on
  Gigantopithecus, 275;
  Nootka Notes, 155; on
  Ostman story, 102–3,
  117, 124–25; on stack-
  ing rocks, 255; taxider-
  mist interview, 75–76
Grenville, Bruce, 142–43
Gross, Harvey, 174, 198
Haas, George, 34
Halpin, Marjorie, 211
Hamlin, Dick, 243
Hartson, Tamara, 168
Hawkins, Tony, 61–62
Hertel, Lasse, 211
Heuvelmans, Bernard,
  9–10
Hewitt, Stan, 170

Hewkin, Jim, 229
Highpine, Gail, 9
Hilson, Stephen E., 124
Hixson, J.E., 274
Howard, Frank E., 101–2
Huff, J.W., 34–35
Hunkeler, C., 274
Jackson, A.H. (Handlogger), 192–93
Jackson, Al, 27, 31–32, 89, 123–24, 196, 198, 221–23, 226, 249, 257
Johnson, Doug, 24–25
Johnson, Steve, 46
Johnson, Warren, 210–12
Johnstone, Bruce, Sr., 194–97, 237–38, 252–54, 265–67, 279
Joseph, Robert, 142–43
Kelley, George, 288
Kirlin, R.L., 211
Krantz, Grover, Ph.D., 12, 55, 75, 167, 271, 287–88
Kristovich, John, 191, 203
Kristovich, Patty, Jr., 197–98
Lauth, Jake, 205–6, 223, 257–58
Lawick-Goodall, Jane van, 271, 273
Lively, Frank, 266
Lorenzetto, Adeline, 158
Mack, Clayton, 98–101, 172
Mariano Mozina, Jose, 154–55
McIlwraith, T.W., 147–49
McKay, Jack, 192
McNair, Peter, 142–43
Meldrum, Jeff, 167, 182, 280, 287–88
Milne, John, 263
Morgan, Jay, 62
Morrison, Dale, 267–68
"Muchalat Harry," 125
Muench, Eric, 239–42
Mulder, Victor, 228–29
Muzzana, Jodi, 189
Nanez, James, 75, 255–56
Napier, John, Ph.D., 14
Needham, Kelly, 246
New Alaskan (newspaper), 144
Nix, Haylee, 81
O'Clair, Rita M., 93
Omacht, Harold, 199, 239
Ostman, Albert, 102–18, 279
Peterson, Andrew, 288
Peyton, Leonard J., 210
Quadra, Bodega y, 154–55
Rohner, R.P. and E.C., 152
Ryan, Charles, 225
Sanderson, Ivan T., 10, 102, 126, 283
Sanderson, Ray, 142
Sanderson, Rob, 202
Sapir, Edward, 155
Schaller, George, 14–15, 246
Shackley, Myra, 272
Sheldon, Ian, 168
Shelton, Rob, 203–4
Shirley, Bruce, 27, 176
Shirley, Scott, 27
Smith, Kenny, 208–9
Sowers, Eric, 62–63
Steller, Georg, 61
Stockli, Jack, 208
Stockli, Rod, 199, 254
Strassenburgh, Gordon, 278
Strough, Mike, 79–80
Struhsaker, T.T., 274
Suttles, Wayne, 157–60, 162
Suzuki, A., 272
Swadesh, Morris, 155
Swanton, John R., 139, 223
Thomas, Larry, 43, 223
Thommassen, Harvey, 98
Timmerman, Dave, 207
Titmus, Bob, 10, 33
Trayler, Chuck, 170
Wallace, Dexter, 142
Wills, Terry, 85–86
Wilson, Iris Higbie, 155

 View all HANCOCK HOUSE titles at **www.hancockhouse.com**

HANCOCK HOUSE *mysteries of the Northwest*

The Best of Sasquatch Bigfoot
John Green
0-88839-546-9
8½ x 11, sc, 144 pages

Sasquatch: The Apes Among Us
John Green
0-88839-123-4
5½ x 8½, sc, 492 pages

Meet the Sasquatch
Christopher Murphy, John Green, Thomas Steenburg
0-88839-573-6
8½ x 11, sc, 240 pages

Bigfoot Encounters in Ohio
C. Murphy, J. Cook, G. Clappison
0-88839-607-4
5½ x 8½, sc, 152 pages

The Bigfoot Film Controversy
Roger Patterson, Christopher Murphy
0-88839-581-7
5½ x 8½, sc, 240 pages

Sasquatch Bigfoot
Thomas Steenburg
0-88839-312-1
5½ x 8½, sc, 128 pages

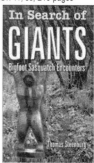

In Search of Giants
Thomas Steenburg
0-88839-446-2
5½ x 8½, sc, 256 pages

The Locals
Thom Powell
0-88839-552-3
5½ x 8½, sc, 272 pages

Shadows of Existence
Matthew A. Bille
0-88839-612-0
5½ x 8½, sc, 320 pages

Rumours of Existence
Matthew A. Bille
0-88839-335-0
5½ x 8½, sc, 192 pages

Strange Northwest
Chris Bader
0-88839-359-8
5½ x 8½, sc, 144 pages

Made in the USA
Columbia, SC
12 December 2024